The Secret Lives of Birds

The Secret Lives of Birds

PIERRE GINGRAS

KEY PORTER BOOKS

First published in Quebec in 1995 by Le Jour, a division of the Sogides group.

Canadian Cataloguing in Publication Data

Gingras, Pierre, 1947–
   The secret lives of birds

Translation of: Secrets d'oiseaux.
ISBN 1-55013-821-9

1. Birds – Behavior.   2. Birds – Food.   I. Feldstein, Peter, 1962–   .
II. Title.

QL676.G5613 1996   598.251   C96-932126-0

The publisher gratefully acknowledges the assistance of the Canada Council and the Ontario Arts Council.

The illustrations used in this book were taken from *Animals* (New York: Dover Publications, 1979), a pictorial archive of illustrations from nineteenth-century sources.

Key Porter Books Limited
70 The Esplanade
Toronto, Ontario
Canada M5E 1R2

Distributed in the United States by Firefly Books

Printed and bound in Canada

97 98 99 00 6 5 4 3 2

Contents

Acknowledgments

This book could not have been published without the assistance of many people. I would like to thank the specialists who, over the years, have steadily and patiently answered my many questions; in particular Henri Ouellet, ornithologist and researcher emeritus at the Canadian Museum of Nature and associate professor at the Université de Montréal; Yves Aubry, co-editor of *The Breeding Birds of Quebec*; Normand David, director of the Association québécoise des groupes d'ornithologues; and Pierre Bannon, who is in charge of compiling observations for the Quebec Society for the Protection of Birds. Mr. Ouellet also revised the final version of the manuscript.

Many thanks, also, to my employer, the daily newspaper *La Presse*, which has been publishing my birding column, "À tire d'ailes," since the spring of 1987, as well as to the readers, who have contributed in many ways.

Finally, special thanks to my son, Frédéric, and my wife, Martha, who also helped revise the manuscript.

Could Cupid Really Fly?

As classically depicted, the dreamily beautiful Cupid would not have been able to fly. With his chubby body and minuscule wings, the Roman god of love would have been incapable of, for instance, the hovering flight of a hummingbird. The Romans who conceived of him knew nothing about the inviolable laws of physics and the biological dictates of flying beings.

It is not simply that Cupid's wings are too small. Even if they were bigger, his lack of a keel, the large bony ridge of the sternum in birds that serves as a stable point of attachment for the flight muscles, would surely have kept him grounded.

THE SECRET LIVES OF BIRDS

What's more, his body is not streamlined, and his center of gravity is poorly located with respect to the position of his wings.

Even if, by some miracle, Cupid managed to fly, his body wouldn't be able to dissipate the heat produced by his efforts. As his temperature rose, his flight attempt would come to an end. In addition, being featherless, the god of love would be at the mercy of storms, the sun and parasites.

Of course, Cupid is merely a mythical creature created by humans who have always dreamed of flying like the birds. In another myth, Icarus managed to escape the Minotaur thanks to the wings stuck to his arms. Unfortunately, they melted under the sun, and the would-be flyer fell into the sea.

It would take several thousand years before human beings would learn how to fly like birds and join them in the skies. It was on October 9, 1890, that Clément Ader launched us into the age of aviation by flying dozens of feet over the ground in a strange vehicle driven by a propeller with feather-shaped blades.

Air Traffic Control

A study of the aerodynamics of birds' wings shows that if Cupid ever hoped to be able to shoot his arrows out of the sky, he would need a wingspan comparable to that of an albatross, a Canada Goose or a Common Loon. For an animal to fly, the surface area of its wings must be big enough to support its weight. The largest wings in the animal world—up to 12 feet (3.5 m) long by 9 inches (23 cm) wide—belong to the Royal Albatross. The average weight of this pelagic, or ocean-going, species is 15 to 20 pounds (7 to 9 kg), although some birds can weigh up to 26 pounds (12 kg). The wing loading, or amount of weight borne by each square centimeter of wing surface, is 1.7 g/cm^2.

This albatross remains at sea for weeks at a time, staying aloft by riding the wind and air currents. It drops onto the water only to catch its prey and eat, or to rest on windless days. Its only sojourn on dry land occurs once every 18 months, when it returns to breed.

At the other extreme, the wing loading of the Barn Swallow is 0.16 g/cm². In passerines such as North American sparrows and Old World warblers it is 0.1 to 0.2 g/cm². In the Canada Goose, it is 2 g/cm².

THE WINGS OF HEAVEN

Could the angels, those messengers of heaven, have done a better job of flying than Cupid?

In Nativity scenes, a plaster angel with long, broad wings attached to its shoulders and folded on its back can sometimes be seen bowing its head while receiving alms. Though impressive-looking, these wings would never have been sufficient for flight. Judging by the considerable size (compared with that of a human) of angels depicted in religious works, they would have needed huge wings in order to fly, beating all wingspan records.

But where flight is concerned, reality can be stranger than fiction. Phenomenal flying creatures existed during the age of the dinosaurs. Cretaceous reptiles had the largest wings ever recorded for any living being. Movies give us spectacular images of these flying animals, but they often disregard the most recent paleontological discoveries.

Despite its fearsome appearance, with its long bill, innumerable teeth and prominent skull, the famed pterodactyl was probably less impressive in real life than as re-created on the silver screen. Some scientists believe it was no larger than a pigeon or a hen; others think it resembled a large raptor.

In Texas in 1975, and five years later in Dinosaur Provincial Park in southern Alberta, Canada, researchers found the fossilized bones of a fantastic, long-billed flying reptile. *Quetzalcoatlus* was human sized, but its bones contained air-filled recesses like those of birds. As a result, it weighed only one-fourth as much as a human. Its batlike, membranous wings were up to 50 feet (15 m) long. All in all, scientists say, the animal must have looked like a small airplane.

Dale Russell, the paleontologist who made the Canadian discovery, notes that, although the animal was rare, it ranged worldwide, as demonstrated by the fossil remains found in Israel and elsewhere.

THE FEATHERED "LIZARD"

Birds are not, of course, the only creatures that fly. Insects were up in the air long before them, and at least one species, the Monarch Butterfly, goes on a spectacular annual migration. In early fall, it flies south from northern regions to winter in the mountains of central Mexico, a distance of some 3,400 miles (5500 km). After spending the winter in a state of torpor, the butterfly gradually returns north, breeding along the way. Its progeny complete the northbound migration. The journey is a feat many birds would be incapable of.

The Archaeopteryx: *A milestone in bird evolution.*

The flying reptiles disappeared long ago without leaving any winged descendents. Rounding out the present-day flyers are the bats, which belong to the class Mammalia.

Certain birds, such as penguins, ostriches and kiwis, are entirely flightless; others make the equivalent of a trip around the world every year. However, all birds have one characteristic found nowhere else in the animal kingdom: they have feathers. Yet the composition of feathers is similar to that of human hair or reptilian scales.

Birds as we know them today appeared about 60 million years ago, after a slow process of evolution. But the emergence of the first feathered animals, representing the evolutionary link between land reptiles and modern birds, dates back even further—to the Late Jurassic period before the appearance of certain large dinosaurs, some 140 to 150 million years ago.

Archaeopteryx, considered to be the first fossil bird, was discovered in 1861 in a Bavarian limestone quarry. During the animal's lifetime, the region basked in a tropical climate. No larger than a pigeon or a crow, the bird was covered with feathers like those of modern birds, and its wings resembled those of game birds (the group that includes chickens, pheasants and grouse). But *Archaeopteryx* also had reptilian features that are not found in later bird fossils, including a rigid, feathered, lizardlike tail composed of 20 vertebrae, and teeth set into both jaws of its horny beak.

The shape of its legs shows that it could perch and that it had claws for climbing. As well, the bird seems to have been a good runner. Using the tips of its wings, it could grab onto vegetation and climb branches.

Morphological features suggest that *Archaeopteryx* was incapable of continuous flight. It lacked the main muscles that power

THE SECRET LIVES OF BIRDS

birds' wings, as well as the strong sternum that flight muscles are attached to. Since its bones do not seem to have been hollow, its skeleton was probably quite heavy.

Some authorities believe the animal could fly for short distances; others think it used its wings only to hop from branch to branch, and that it could not take off from the ground.

LEARNING TO FLY

On the whole, then, opinion in the scientific community is not unanimous on *Archaeopteryx*, although no one doubts that it is an important stage in the evolution of reptiles into birds. Moreover, some paleontologists believe that other fossil animals may also have been the ancestors of birds, particularly one small crocodilian with long scales that, according to this theory, evolved into feathers.

The scientific debate over *Archaeopteryx* inevitably brings up the following question: how did birds learn to fly?

One theory is that, originally, feathered lizards made use of their wings only to glide from branch to branch. Over time, the wing bones, sternum and muscles of these prehistoric animals developed enough to support long-distance flight.

Another theory holds that *Archaeopteryx* was a ground dweller whose wings helped it spring upon and capture its prey. Gradually, over the millennia, its muscular structure developed the capacity for true flight.

Archaeopteryx is merely one stage in a long process of evolution, and many bird fossils have been discovered subsequently. Some were seven- to 10-foot (2- to 3-m) flightless giants; others, such as *Teratornis incredibilis*, a sort of giant vulture with a 16-foot

(5-m) wingspan, were powerful flyers. Discovered in California, this fossil bird cruised the skies about 15,000 years ago.

Passerines on Top

Impressive as these massive birds were, it is the small, compact passerines that have come to represent the acme of avian evolution.

Until recently, the first passerines were thought to have appeared about 25 million years ago. But in March 1995 an Australian zoologist discovered a fossil that led him to push back their origins by 30 million years.

The passerines are by far the largest order of birds, with some 5,700 of the world's 9,700 species. Familiar examples of passerines are sparrows, Old World and New World warblers, swallows, weavers, starlings and crows.

Passerines like this Kingfisher represent
the acme of avian evolution.

A Pound of Feathers

Birds are the only living creatures with feathers. But contrary to what you might think, they are not uniformly distributed over their bodies.

The Whistling Swan is the species with the most feathers—more than 25,000, with 80 percent of them concentrated on its head and neck. Among the birds with the fewest feathers are hummingbirds, with just 900. A Barn Swallow has about 1,400 feathers, a Brown-headed Cowbird about 3,800 and a Green-winged Teal about 14,000.

In general, small birds are more densely feathered than big ones. The fundamental biological explanation for this is that the smaller the animal, the more energy it expends, relative to its weight, on maintaining its body temperature.

A bird's feathers normally weigh two to three times more than its skeleton. For example, the 7,000 feathers of a Bald Eagle make up 17 percent of its body weight, compared with 7 percent for its bones. Most songbirds and small passerines have between 2,000 and 4,000 feathers, 35 percent of which are located on the head and neck. The plumage of these birds is fuller in winter than in summer.

BARE PATCHES

Although feathers look as though they are uniformly distributed over a bird's body, there are patches where none grow at all. In fact, feathers grow out of highly localized sites, usually eight in number, known as *pterylae*.

The bare patches, such as those found on the upper thighs, are called *apteria*. These areas are covered by a fine down that enables the animal to dissipate heat. The only birds lacking apteria are the penguins, the small African mousebirds and the screamers, turkey-sized birds of South America.

Feathers are not all the same. They have differing forms, depending on where they are located on the body. The large wing feathers, numbering nine to 12, are called *primary* and *secondary remiges*. Tail feathers are called *rectrices*. Most of the body is covered by small *contour feathers*. Insulation is provided by *semiplumes* and *down*. There are also *filoplumes*, hairlike feathers that inform the bird, through tactile stimuli, of the position of the other feathers that play an active role in flight.

pterylae

apteria

Feathers are gathered in areas called pterylae.

The bristles around the bills of nighthawks and other insectivores, as well as those around the eyes of cuckoos, are feathers, too. Their function is to protect those parts of the body, while giving the birds tactile feedback on the prey they are preparing to swallow.

DEADLY FEATHERS

Powder down, delicate feathers producing a fine waxy powder that keeps the plumage healthy, is also present in most birds, but it is particularly abundant on certain waders such as herons and bitterns. The substance sometimes lends a grayish hue to the plumage. In the Great Blue Heron, whose diet consists mainly of fish, this powder apparently helps remove any fish mucus that adheres to its feathers while it is feeding.

Most feathers are centered around a stiff shaft with a hollow base called a *rachis*. The down growing on the bottom of the shaft is important as insulation.

Feathers grow from follicles under the skin. As with mammal hair, once a feather has stopped growing, the portion of it extending outside the body dies. Although feathers cannot change color, their hues can fade through wear and exposure to the elements.

Growing from both sides of the rachis are the *barbs*, which are further divided into interlocking *barbules*. A large pigeon feather has 600 pairs of barbs. They form a dense network that appears unbroken to the naked eye. The rectrices of large waders may have up to one million barbules.

Undoubtedly, the most remarkable feathers are those of the pitohuis of New Guinea, which contain a deadly poison. In

THE SECRET LIVES OF BIRDS

Ornithology, author Frank B. Gill states that this 1992 discovery raises many questions.

Local people have long known that the flesh of this bird would make them sick if not prepared in a special way. The poison in question is an alkaloid with a chemical composition similar to that found in certain South American frogs, which has been used in poison arrows. One fascinating mystery is why the bird does not poison itself when preening its feathers.

SCARED FEATHERLESS

Feathers are made of keratin, a protein substance that is also a basic component of hair, fingernails, fur, wool and animal hooves. Feathers are replaced a few at a time; old ones fall out and new ones grow in.

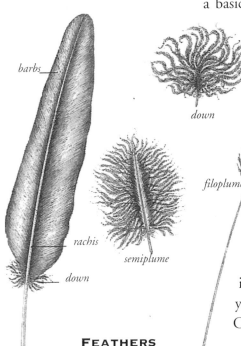

barbs

down

rachis

down

semiplume

filoplume

FEATHERS

Renewal of outer layers is one of nature's constants. We humans lose our hair and are constantly growing new layers of skin. Insects shed their exoskeletons; plants drop their leaves. Birds, for their part, molt.

This phenomenon is universal in birds and occurs at least once a year in most temperate-zone species. Certain sparrows, for example, undergo a complete molt twice a year. Ptarmigans take on three successive plumages during the year: brown and white in the fall, completely white in winter, and brown and white again in spring. These three molts make for perfect camouflage during each season of the year.

Birds molt because their feathers become worn and no longer play their protective role. Harsh weather, ultraviolet light and wear during the nesting period are some factors that affect plumage.

The first molt of a bird's life occurs while it is still a nestling, when its light down is gradually replaced by a warmer coat. The down is pushed out by the growing feathers, which are more pliable in young birds than in adults. Despite their annual molt, gulls keep their juvenile markings until they reach breeding age, two, three, even four years after fledging, depending on the species.

Many passerines molt at the end of the summer once their young have begun to fly. However, the Snow Bunting molts even earlier, due to the shorter summer at its northern breeding grounds.

Molting usually occurs symmetrically, with feathers gradually falling out in equal numbers from both wings so that the bird's balance remains unaffected. Most birds continue to fly during this period, although their flight is less efficient and requires more energy. As well, the molting process takes place at a time of year when food resources are at their most abundant. In the Chaffinch, energy expenditures increase 25 percent during this period.

Geese and ducks molt at the end of the breeding season. The partial molt of ducks results in a drab plumage called an *eclipse plumage*. The magnificent breeding colors of ducks reappear during the winter.

Molting is a critical time for geese, ducks, swans, loons, grebes, flamingos and several species of cranes because they lose their ability to fly and become more vulnerable to predators. As a result, they are more cautious during this period. Male ducks leave their nesting territories and gather on large lakes, for example. Females, on the other hand, molt in August, when their ducklings are almost fully grown.

Ducks raising a second brood after the first has been lost to a predator or flooding may molt very late in the season. They will be true "sitting ducks" when the hunting season opens in Canada at the end of September!

In geese, males and females molt at the same time, when their young are about ready to fly.

Paul Géroudet, in *Grands échassiers, gallinacés, râles d'Europe*, discusses another molt, which he terms the "fright molt." Some game birds such as the Willow Ptarmigan, Hazel Grouse and Capercaillie will drop some of their feathers when startled. This rather unusual reaction could distract the predator, enabling the bird to escape.

TRUE (AND FALSE) COLORS

Feathers are at their most colorful during the breeding season and the weeks leading up to it. The males of a large number of species are "better dressed" than the females, an aspect of a larger phenomenon known as *sexual dimorphism*.

Some examples of particularly striking birds are ibises, parrots, Northern Cardinal and Eurasian Golden-Oriole. However, the drab plumage of other birds makes them practically invisible in their environment, a phenomenon known as *protective coloration*. You can stare straight at an American Woodcock or Whip-poor-will sitting on the ground without seeing it.

Feather color, like human skin and hair color, is often the effect of pigmentation. However, in birds, some colors are created by an entirely different phenomenon. Most blues and greens, for example, are the result of a sort of optical illusion. And the iridescence found in hummingbirds, the European Starling, ducks and others is caused by the structure and angle of certain feather components.

It is the reflection of light that allows us to perceive birds' colors. When an object reflects all wavelengths of light, it is completely white. When it absorbs all light, it is black. This is why a black coating absorbs much more heat than a white one. A red object is one that absorbs all colors of the visible spectrum except red.

LOOKING A BIT PALE

In birds, the color of the feathers, bill, legs and skin are due, most of the time, to innumerable pigments that reflect only certain colors.

The commonest pigment in feathers is melanin. It is the basis of colors ranging from black to gray to brown. Melanin can combine with other pigments to produce pinks and yellows.

Carotene is the most widespread pigment in nature. However, unlike other pigments, it cannot be synthesized by animals, and birds must eat products that contain it. This causes headaches for zoo keepers trying to provide a carotene diet for each species of red bird. A good example is the American

Flamingo. Its feathers will molt to white if its diet is not sufficiently rich in carotene.

A deficiency or an excess of certain pigments can produce unusual color variations. Cases of melanism (black) are infrequently observed but are common enough among American Robins. Some are so black that they can easily be confused with their European cousin, the Blackbird. Conversely, cases of partial or complete albinism, caused by the absence of pigment, are quite common. Completely white American Crows have been spotted. Another unusual case is the Evening Grosbeak, which has been observed without its characteristic black feathers.

BLUES AND GREENS

Blues and greens, among the most spectacular of hues, are generally not produced by pigments. The bright green of parrots, parakeets and hummingbirds as well as the blue of the Eastern Bluebird, the Blue Jay and the Blue Rock Thrush are due to an entirely different physical phenomenon. Blue and, indirectly, most bright greens are created by the structure of the feather itself. The barbs are equipped with minute air pockets *(vacuoles)* that reflect blue wavelengths of the visible spectrum. Since

The green color of touracos is caused by a rare pigment called turacoverdin.

the remainder of the feather possesses a relatively dense layer of melanin, all other colors of the spectrum are absorbed.

The origin of most greens is even more remarkable. In fact, it turns out that most beautiful green parakeets are actually blue! The green results from the blue being reflected by the vacuoles through a fine layer of yellow carotene, so that the secondary color our eye perceives is green. There are a few exceptions, such as in the magnificent African touracos, whose green color is produced by a very rare pigment called *turacoverdin*.

The iridescence that creates green or blue speckles is primarily the result of the observer's position and the arrangement of the feather barbules. Depending on the angle of reflection, the European Starling's plumage and the neck or wings of certain ducks can take on a variety of shimmering colors.

In 1994, it was found that the green *caruncles* (fleshy outgrowths) above the eyes of the male Velvet Asity of Madagascar are of even more complex origin. Frank B. Gill indicates that their color could be due to light being reflected onto protein fibers contained in the caruncle during the breeding period.

The Importance of
Good Hygiene

Birds devote much time to their hygiene, preening their feathers continually and bathing regularly. Many birds spread their wings in the sun. Others prefer a good dust bath, smear themselves with an ant ointment or enjoy a beer massage.

Birds need not look long or far to find the beauty oil they need for their hygiene. It is just a bill's length away, secreted by the *uropygial gland* on their rump.

If you take a close look at a chicken before cooking it, you will notice this small orifice on top of the rump. Usually removed at the slaughterhouse, this gland secrets an oil rich in fatty acids and water. Birds dip their bills into this oily substance up to 10 or 12 times a day and spread it on their feathers.

A Vitamin D Supplement

This oil helps keep the plumage healthy and removes dust, as well as keeping the parasite population down. It also keeps the feathers waterproof (for insulation, particularly important in ducks) and flexible (for flight). In a chemical reaction induced by sunlight, the liquid secreted by the uropygial gland is partially converted to vitamin D. This vitamin supplement is absorbed by the animal when it preens its feathers.

The uropygial gland is lacking in the Eurasian Green Woodpecker.

In cormorants, the secretions by themselves are not enough to keep the feathers waterproof. That is why these birds are often seen with their wings spread out to dry after the birds emerge from the water. The uropygial gland is lacking in ratites (emus, ostriches and others), pigeons, woodpeckers and parrots.

Birds also preen themselves to get rid of the many parasites (including more than 200 species of lice, fleas and ticks) found on their feathers and skin. Too many parasites can seriously affect a bird's health, and nest infestations can be particularly harmful to chicks.

Preening can be a social activity. Herons, albatrosses, penguins, whistling-ducks, pigeons, parrots and Eurasian Bullfinches have been known to preen their fellows. In captivity, this behavior is even observed between birds of different species. Certain cage birds, such as parrots, also like their heads to be scratched.

Goatsuckers, herons, the Barn Owl, frigatebirds and dippers (small birds that hunt their prey underwater on stream bottoms) all have a toothed middle toe—their own personal "comb"—which they use to preen their head, neck, throat and bristles. The spaces between the teeth of this "comb" are just wide enough to allow the feather barbs to pass through so that downy debris and dust are removed.

A SLUSH BATH

Since birds have no sweat glands, they eliminate most of their water through respiration and excretion. Their digestive systems reabsorb and reuse a large amount of water, making their excreta relatively solid.

Birds love to bathe in puddles. Not only does this keep them cool, but it also helps them maintain their plumage. Most birds, with the exception of game birds, woodpeckers and a few others, love water and will even bathe in winter if an opportunity arises.

Hummingbirds, like many familiar birds, appreciate a shower under a lawn sprinkler.

Large birds such as eagles, falcons, crows, even owls, particularly like to bathe along stream banks and in swamps. While standing in the water, they will wash themselves slowly and stay still for long periods of time. Then, with a few sharp wingbeats, they dry themselves off.

Bathing behavior in smaller birds has a characteristic pattern. First, they take a few steps into a few inches of water. Next they thrust in their heads and quickly lift them out again. Then they crouch for a few moments, beating their wings while cocking their tails. When the bath is over, they shake themselves off vigorously, retire to a quiet place and carefully preen themselves. While bathing, smaller birds are always wary, since they know being wet makes takeoff more difficult, and hence makes them more vulnerable to predatory birds.

Birds can make do without a puddle for their baths. Some fly through wet shrubs to get a shower from the wet leaves. Others use the dew-soaked grass in the morning. Parrots and woodpeckers wait for a good storm. Some flycatchers bathe by diving into the water.

Bird bathing sites are found in the most unlikely places. For example, birds have been observed soaking themselves in tiny pools in a small cavity of a tree trunk, or in wet snow. During northern winters—even at −4°F (−20°C!)—several species will use man-made birdbaths that are kept from freezing by a heating element. Mourning Doves, American Goldfinches, House Sparrows, European Starlings, even American Robins wintering in the North, are particularly given to this activity.

THE BENEFITS OF DUST

Some birds love water—others prefer dust.

In the barnyard, chickens and turkeys can be observed wallowing in the dirt—a practice that seems to bear little resemblance to a conventional bath.

House sparrows, too, often scrabble around in the dust. They crouch in a hollow, claws extended, wings spread, beating up dust into their plumage. When the dust bath is over, they shake themselves off and then preen their feathers.

Dust baths are a frequent activity of many game birds, larks, wrens and raptors, including falcons and some owls. The ostriches of Africa, along with their Australian cousins, the emus, love to roll around in the dirt.

Why birds do this is still unknown, but it is thought that dust helps keep their plumage healthy. It removes some of the oil and dead skin cells that build up on the body, and it dries up moisture accumulating in the feathers, making for better insulation.

As well, the bath helps realign the barbs on either side of the feathers' main shaft. The dust is thought to flush out certain skin and feather parasites for whom drier skin and better-aligned plumage are a less accessible and less attractive habitat.

Another element of bird hygiene is sun bathing. Here birds spread their wings in cool weather, even in hot temperatures, to better absorb the heat of the sun.

Sun baths apparently stimulate the production of vitamin D in the feathers and help them regain their shape after vigorous flight. Heat also forces skin parasites to move about more, making them easier to capture. A bird can then rid itself of these undesirable and otherwise hard-to-reach hangers-on with its bill or claws.

AN ANT OINTMENT

Ants are a part of many birds' diet, but they are also sometimes used as an "ointment." A bird will crush an ant with its claws, mix it with saliva and spread the ointment on the top of its head and under the feathers at the end of its wings, normally leaving the breast untouched. The organic fluids contained in ant flesh are composed of essential oils that the bird uses to preen its feathers—a sort of beauty cream.

At least 200 bird species, particularly passerines, use ants in this way. The ritual can last for a quite some time. In some cases, the bird makes a pile of its victims, then devours them or simply leaves them behind.

Some birds will simply plant themselves on an anthill, their wings pointing forward and draped on the ground, and allow the ants to crawl up into their feathers. The ants have no pincers with which to defend themselves, but they do secrete formic acid as a means of defense. The anal gland of these insects also produces another substance. Its odor is repellent, and, like formic acid, it has insect-repellent properties. As John K. Terres writes in *The Audubon Society Encyclopedia of North American Birds*, it has been shown that a solution of this acid can kill certain feather mites. Thus, it is believed that these "ant baths" serve to eliminate parasites.

Among the species exhibiting this behavior are the Blue Jay, Eurasian Jay, Eastern Bluebird, Blackbird, crows, certain woodpeckers, Evening Grosbeak and House Sparrow.

When ants are lacking, several species turn to any of more than 40 substitutes, ranging from naphthalene to orange juice and including vinegar, hot chocolate, beer, cigarette butts or other insects. Ostriches are known to use bees for this purpose.

There is one reported case in the scientific literature of a Black-billed Magpie in England that enjoyed having lit cigarettes held against its feathers because of the heat. How the feathers held up under this stress is not mentioned!

The most amazing anecdote comes from Great Britain, where a tame magpie flew to its owner's shoulder with an ant, dipped the insect into the smoking ash in his owner's pipe, then smeared itself with the hot ointment.

Blue Jays love to wallow in ant nests.

Up, Up and Away

Flight is a very complex activity. As in airplanes, flight in birds involves working constantly against gravity. That is, a bird's wing functions to provide both lift and forward propulsion.

The large feathers at the wingtips, known as *primaries*, act like the propellers in an airplane. On the upstroke in flapping flight, the wing is slightly tilted, with the "wrist" (the part of the wing bearing the primaries) held vertical and the primaries partially separated to decrease resistance. At the top of the stroke, the wrist flaps abruptly up and out, assuming a horizontal position from which to push the air backward and toward the ground. Contrary to what you might think, birds do not flap their wings straight up and down.

In exact analogy to the motion of an airplane wing, the displacement of air toward the more rigid, incurved middle of the wing provides the lift a bird needs to gain altitude. This physical principle can be illustrated in a very simple way by slightly folding a piece of paper and blowing on it. If you try this, you'll see that the top of the paper rises.

The greater the speed, the greater the lift and the more altitude the bird will gain. Birds are constantly adjusting the angle of their wings, depending on whether they want to gain, lose or maintain their altitude. The tail also plays a role in steering, but each wing can act independently to produce the desired changes in direction.

Birds must carefully control the landing operation to avoid crashing. Using its *alula*, a small feathered projection at the bend of the wing farthest out from the body, a bird can better control air displacement and avoid dropping too abruptly.

HOVERCRAFT

The complex flight apparatus in birds is driven by the coordinated action of about 50 different muscles. It has been calculated that the European Starling beats its wings on average 4.3 times a second in normal flight. The figure is 3.2 for the Common Pheasant, 4.9 for the American Goldfinch, 2.4 for the American

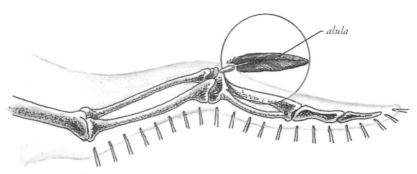

alula

DETAILS OF A WING AND ITS ALULA

Kestrel and 2.0 for the Black Duck, a species closely related to the Mallard. The Ruby-throated Hummingbird female and male beat their wings 53 and 70 times a second, respectively.

In general, the bigger a bird's wings, the slower the wingbeat needed to keep it in flight. Some species, including large raptors and Old and New World vultures, soar on rising air currents. The air lifts the wing mechanically and the bird gradually gains elevation. At the top of each thermal, it glides down to another one and begins the cycle again.

HOW TO MAKE AN OSTRICH FLY

Each species has its own characteristics in terms of speed, power and maneuverability, depending on the size of the wings and the weight they bear.

Game birds such as pheasants and grouse have short, broad wings that enable them to take off vertically in the forest— where the runway is usually very short!

To fly, an ostrich would have to reach a ground speed of 100 mph (160 km/h).

Small passerines can perform stunning acrobatics and effortlessly make their way through dense brush. At the other extreme, albatrosses, with their large wings and heavy bodies, have great difficulty getting into the air. Without a wind, they are quite simply stuck on the ground—or, more likely, on the water. Obviously, these birds are incapable of the aerial antics of, say, a chickadee.

Loons, guillemots, puffins, Razorbills and diving ducks such as scaup and goldeneye have relatively great wing loading. For these species, takeoff is a laborious operation that involves running across the surface of the water while rapidly beating the wings.

Ostriches, which weigh about 285 pounds (130 kg), would have to reach a ground speed of 100 mph (160 km/h)—which of course they cannot—in order to become airborne. Other flightless birds include emus, kiwis, penguins and the Flightless Cormorant of the Galápagos Islands. For unknown evolutionary reasons, the wing bones and flight muscles in these species have atrophied over time. In penguins, the wings are used for swimming rather than flying.

GET IN FORMATION

Of all North American migrations, the most spectacular is probably that of the Canada Goose. The long lines of squawking geese that darken the sky mark the arrival of spring or the end of autumn. Most of the time the geese fly in a very orderly fashion in V-shaped flocks. But quite commonly, they will travel in a diagonal line up to several hundred feet long.

Other birds that fly in formation are cormorants, Common Eider, many other duck species and other geese, including the Snow Goose, Brant and Graylag Goose. However, in these species, the formations are not as tight and well organized.

Flying in close formation reduces energy expenditure. With

THE SECRET LIVES OF BIRDS

each wingbeat, the bird displaces air toward the tips of its primaries, creating an upswelling vortex to its rear that buoys the next bird in line. The suction is similar to that experienced by a car driver behind a large, fast-moving truck.

While flying in formation, each bird positions itself slightly higher than the bird in front of it in order to take maximum advantage of the vortex. Flight in V formation and diagonal lines gives geese a considerable advantage over birds flying alone—theoretically, they gain 71 percent more distance for the same energy expenditure.

The energy expenditure of the leader is compensated for by the energy savings of the followers. So that all birds get the benefits of flying in formation, the bird at the head of the flock regularly switches positions with one of the followers.

As birds have evolved, most species have exhibited flocking behavior. This behavior is now genetically coded in many species. It is believed that the reason small birds do not fly in flocks is that, because of their size, they cannot displace enough air to assist their fellows.

SPEED DEMONS

Flying is a very demanding activity, and birds try to reduce their energy expenditures as much as possible. Cruising speed is usually fastest during migration. For example, a Common Swift will fly about 25 mph (40 km/h) during migration, but only 14 mph (23 km/h) at other times. The speeds at which birds fly vary according to species, but also depending on the circumstances. A bird trying to escape a predator will fly much faster than one merely hopping from perch to perch. Ruby-throated Hummingbirds, which hover to feed on flowers, can also fly more than 50 mph (80 km/h). Wind speed is another key factor that can affect a bird's flight speed.

However, measuring flight speeds accurately is difficult. Many improbable figures have been published, some of them contradictory. One report, for example, gave the European Starling's cruising speed as about 21 mph (34 km/h); another mentioned 37 to 48 mph (60 to 78 km/h); while a top speed of 56 mph (90 km/h) has been recorded.

According to a number of studies, the Peregrine Falcon reaches speeds of 37 to 62 mph (60 to 100 km/h) in normal flight. In England, this falcon—considered the fastest bird in the world—was clocked at 62 mph (99 km/h). In the time of Henry IV, a falcon that escaped from Fontainebleau, in France, was found the next day on the island of Malta, 1600 miles (2600 km) away, which translates into an average speed of 56 mph (90 km/h).

Also recorded in the scientific literature is the case of a Peregrine Falcon that, while stooping (diving) to catch a prey, passed an airplane that was descending at 173 mph (280 km/h). Researchers have also reported the phenomenal speed of 211 mph (340 km/h) for a White-throated Needletail in India. Other amazing performances reported include those of a Red-breasted Merganser in Alaska,

The Herring Gull's average flying speed is 24 mph (38 km/h).

which took advantage of a 15 mph (25 km/h) wind to travel at 99 mph (160 km/h), and a Green-winged Teal, which flew at 58 mph (94 km/h).

But in general, birds fly much slower, usually between 19 and 37 mph (30 and 60 km/h). Typical speeds are 17 mph (28 km/h) for the House Sparrow, 43 mph (70 km/h) for geese, 47 mph (76 km/h) for the Common Eider, 24 mph (38 km/h) for the Herring Gull and 20 mph (32 km/h) for the Barn Swallow.

FLYING HIGH

Impressive as these speeds are, the altitude reached by some birds, especially during migration, is even more incredible.

Most birds migrating by day to their breeding or wintering grounds maintain an altitude of 500 to 1,150 feet (150 to 350 m) in clear weather. However, in a storm they fly much closer to the ground so that they can take refuge quickly in an emergency. You can see thousands of birds, for example, gathered on a spit of land for days as they await suitable conditions to cross a large body of water.

Nocturnal migrants tend to fly much higher, between 2,000 and 3,300 feet (600 and 1000 m). Large birds, such as ducks and geese, usually fly at a higher altitude than small ones, but there are exceptions. At Cape Cod, radar detected many passerines flying at altitudes from 13,000 to 21,500 feet (4000 to 6500 m).

In North America, the altitude record for night flyers is held by an Evening Grosbeak that was spotted flying over the Rocky Mountains in Colorado at 13,000 feet (4000 m) or higher.

Geese often fly at over 6,000 feet (1800 m) or even 8,000 feet

(2500 m) during migration, while nighthawks can reach 9,000 to 10,500 feet (2800 to 3200 m). Other altitude records are 21,000 feet (6300 m) for the Wallcreeper, a small European bird with pink wings; 6,500 feet (2000 m) for the Rook and 10,000 feet (3000 m) for a number of shorebirds.

MIDAIR COLLISION

The most extraordinary altitude performance of all was recorded over Abidjan, on the Ivory Coast. On November 29, 1975, an airplane collided with a Rüppell's Griffon, an East African species, at 37,900 feet (11,560 m). The plane had to make an emergency landing following the incident. No one was hurt, but the plane's engines were damaged.

In the Himalayas, a flock of Bar-headed Geese was observed on migration as it flew over Mount Makalu (29,000 feet/ 8700 m). In 1924, mountain climbers approaching the summit of Everest were accompanied for some time by a Yellow-billed Chough, a crowlike species also found in the Alps. The bird, which was feeding on scraps of food from the expedition, was identified at 26,500

Rüppell's Griffon can reach an altitude of more than 36,000 feet (11,000 m).

feet (8100 m). Of all birds, this species is believed to be the one that nests at the highest altitude.

A Lammergeier, the largest European scavenger species, was sighted at 25,600 feet (7800 m) on the slopes of Everest. Smaller species are also regularly observed flying over the Alps or the Himalayas on migration. These include curlews and godwits, both of which have been spotted at 20,300 feet (6200 m).

In Europe, crows have been seen at heights ranging from 2,300 to 7,200 feet (700 to 2200 m). Elsewhere, Chimney Swifts have been noted by airplane pilots at that altitude.

In North America, the altitude record for a bird was set by a Mallard on July 9, 1963. The bird collided with a Western Airlines jet at 20,900 feet (6370 m). A study of feathers recovered from the airplane following the collision positively identified the species that caused it.

But how do birds endure the lack of oxygen and the cold at such altitudes? At 20,000 feet (6000 m), where there is half as much oxygen as at sea level, an untrained human being at rest has great difficulty breathing. At over 23,000 feet (7000 m), he or she goes into a coma and quickly dies. This is not the case for certain birds, which are able to draw oxygen from the rarefied air at high altitudes. As well, they are protected from the cold by the heat they generate in flight and the thermal insulation their feathers provide.

A Bald Eagle has about 7,000 feathers, much fewer than the 25,000 found in certain swans. (Page 21)

Partially albino birds are quite common, but completely white ones like this Barn Swallow are very rare. (Page 28)

Preening is a social activity in the Eurasian Bullfinch. (Page 33)

The American Robin will bathe in winter in a birdbath kept above freezing by a heating element. (Page 34)

Widespread in Europe, the Blackbird takes pleasure in an ant bath. (Page 37)

Upon reaching adulthood, this young Peregrine Falcon will be one of the fastest birds in the world. (Page 44)

Golden Eagles attack animals much larger than themselves. In Scandinavia, they are major predators of young caribou. (Page 58)

Certain birds use tools to obtain their food. Green Herons will sometimes toss bait onto the water to attract small fish. (Page 61)

Bills for Every Purpose

Bird bills come in a wide range of sizes and shapes—long, short, flat, curved, delicate or massive—and a chorus of colors. In essence, a bill is just an orifice through which food enters, but its many functions also include capturing prey, preening and nest building.

In geese, ducks, flamingos and shorebirds, the bill is used to filter food out of the water. Shrikes and raptors use their bills to kill prey and tear it to pieces. A seed eater uses its bill to crush seeds with a jackhammerlike motion. When breaking a cherry pit, for example, the Hawfinch, a European species, brings its two mandibles together with a force of 100 pounds (45 kg). Parakeets and the like use their bills for climbing.

If oystercatcher bills act like a pair of pliers to pry open mollusks, the bills of woodpeckers are more like chisels. They are used to dig nest holes in trees or capture insects living in the wood. The woodcock, on the other hand, uses its long, sensitive bill to probe the ground for worms.

Bird bills are composed of keratin, a protein substance also found in feathers and claws. It wears down with use, especially in seed-eating birds that feed on the ground. But nature does its work well, since the bill continually regenerates. Of the two mandibles that form the bill, only the lower one is movable.

PIGEON MILK

Food passes from the mouth into the crop, an organ designed for food storage that is unique to birds. The first stage of digestion takes place in this elastic pouch, which also serves as a sound box in some grouse species. Here, food is moistened and tenderized.

The crop is lacking in some seed-eating passerines, as well as ducks, geese, cranes, Old World warblers and thrushes. However, in many cases these birds can still store food in their expandable esophaguses.

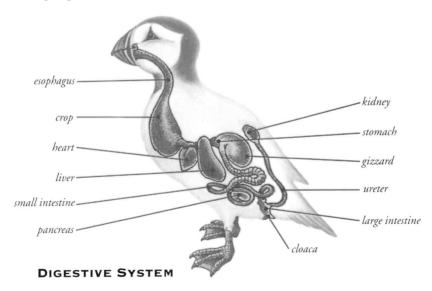

esophagus

crop

heart

liver

small intestine

pancreas

kidney

stomach

gizzard

ureter

large intestine

cloaca

DIGESTIVE SYSTEM

The crop is also well developed in seed eaters such as chickens, grouse and other game birds, as well as pigeons. The crops of pigeons and doves are unique in the bird world. For the three weeks during which these birds raise their young, this organ secrets a thick, rich substance called "pigeon milk." Composed mainly of protein and fat, this liquid is the chicks' only food for the first few days after they hatch. After a while the parents begin to supplement the diet with seeds that have been tenderized in the "milk."

PUT THAT IN YOUR GIZZARD!

After remaining in the crop for some time, the food continues on its way toward a narrow stomach formed out of a recess in the wall of the gizzard. Here, the food starts to be broken down chemically by gastric enzymes. Stomach acidity is very high in carnivorous birds. Terres reports that the Lammergeier, an Old World scavenger, can digest a large vertebra in two days.

Chickadees, House Sparrows, Common Redpolls and finches assiduously hull their seeds, while doves, pigeons and game birds swallow them whole, husk and all. This difference in feeding behavior reflects the different digestive systems in these bird groups. The gizzard of a pigeon performs the same seed-crushing function as a powerful bill in other birds.

The gizzard is to birds what jaws and teeth are to mammals. With its strong muscles, the gizzard acts like a mill, crushing and grinding the food in preparation for the work of the intestinal flora and gastric enzymes at a later stage of digestion.

The power of the gizzard muscles has always fascinated researchers. For instance, the gizzard of a Wood Duck, a widespread

North American species, can purée walnuts. Common Eiders can easily grind shells, particularly those of common blue mussels.

In one seventeenth-century account, a turkey was made to swallow glass balls, lead cubes and a small wooden pyramid. One day later, the glass had been pulverized, the lead flattened and the wood worn down.

In another experiment several decades later, tin-plated tubes were crushed by a turkey gizzard. Obtaining the same result with a pair of pliers would have required a pressure of over 440 pounds (200 kg). The gizzard of this species can grind up steel needles, even scalpel blades.

A GOLD NUGGET

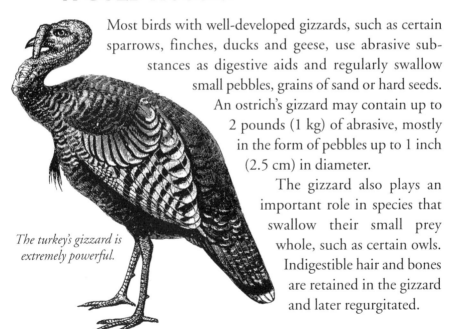

Most birds with well-developed gizzards, such as certain sparrows, finches, ducks and geese, use abrasive substances as digestive aids and regularly swallow small pebbles, grains of sand or hard seeds. An ostrich's gizzard may contain up to 2 pounds (1 kg) of abrasive, mostly in the form of pebbles up to 1 inch (2.5 cm) in diameter.

The gizzard also plays an important role in species that swallow their small prey whole, such as certain owls. Indigestible hair and bones are retained in the gizzard and later regurgitated.

The turkey's gizzard is extremely powerful.

Many unusual objects have been found in bird gizzards. A gold rush was even touched off in Nebraska at the turn of the century when nuggets were discovered in the gizzards of ducks killed during the hunting season. And an emerald mine was reportedly discovered in Burma after a precious stone was found in a pheasant's gizzard.

Unfortunately, this habit of using abrasives has caused the death of millions of wild ducks and geese in North America. These birds sometimes swallow the lead shot discarded by hunters, particularly in shallow marshes. Even at small concentrations, lead is very toxic in wildfowl. It attacks the nervous system and causes paralysis. Just a few tiny pellets can be fatal. For this reason, hunters are now banned from using lead-shot-filled cartridges in the United States, and steel shot is now the norm. Canada is expected to follow suit in 1997.

RAPID-FIRE DIGESTION

Partially digested food leaving the gizzard continues on its way to the small intestine, where nutrients are absorbed. The final stage in digestion occurs in the large intestine, where water is reabsorbed before excretion.

On average, a bird's digestive system is eight times longer than its body. However, this ratio can be much smaller in certain species, such as the Common Swift (3 to 1). At the other extreme, an ostrich's digestive tract is 20 times longer than its body.

Digestion is usually very rapid—about half an hour—but it can also extend over a period of 12 hours in exceptional cases. A notable example of a bird with quick digestion is the Turkey

Vulture, a widespread scavenger in the United States and southern Canada. One Turkey Vulture that swallowed a yard-long snake one-fifth its weight was found to have digested it in 90 minutes.

Although hunting for food is the main preoccupation of birds, the frequency of their meals varies considerably from one species to another. The smallest birds must eat very often due to their extremely high metabolisms. During one English winter, a tit had to eat nearly 90 percent of the time in order to maintain its body temperature. A Willow Ptarmigan, however, can subsist on only one meal a day.

Some birds are remarkable for the variety of their diet and their opportunism. Toucans, for example, eat mainly fruit, but will readily prey on small birds when the opportunity arises.

The Blue Jay and its European counterpart, the Eurasian Jay, are interesting examples of opportunism. Blue Jays will eat practically anything they find. Their habit of storing seeds underground plays a useful role in forest regeneration. They can carry up to five acorns or 14 beechnuts at a time in their expandable crops. In New England, it was calculated that about 50 Blue Jays carried and buried no fewer than 150,000 beechnuts in 23 days.

But Blue Jays also eat all sorts of seeds, berries, insects, spiders, snails, crayfish, small fish, frogs, mice, even bats. As well, they prey on bird nestlings and eggs. The Eurasian Jay's diet is equally varied.

Birds that hide seeds, or, like shrikes, impale their prey on hawthorn needles and eat them later on, are quite adept at relocating their caches. For example, the Tufted Titmouse will store

50,000 seeds in one fall season in as many different spots, and will return to many of them later. The Clark's Nutcracker manages to find pine nuts hidden nine months earlier. These birds use visual cues to home in on their caches.

THE FEATHERED VAMPIRE

The diet of the Yellow-bellied Sapsucker is rather peculiar. A virtual "tree vampire" among birds, it bores a series of equally spaced holes in the bark of a tree and then uses its long tongue, the tip of which is equipped with fine appendages, to suck out the sap. Sapsuckers have been observed feeding on 275 different North American tree species. Small wonder that these birds were once heavily persecuted by woodland owners.

For its "victims," the Yellow-bellied Sapsucker chooses trees with the most sap. The hemorrhage it causes is rarely fatal, but repeated attacks weaken a tree, sometimes making it easy for fatal diseases to enter. The browsed portion also becomes more fragile and can break off in high winds.

The niche occupied by the Yellow-bellied Sapsucker in the Americas is filled, in Europe, by the Three-toed Woodpecker. Oddly, this species rarely attacks trees in North America.

ENERGY CONSCIOUS?

Sapsuckers are not the only animals to feed in this way: insects, other woodpeckers and the Ruby-throated Hummingbird do so, as well.

For hummingbirds, the champion energy consumers among birds, tree sap is a welcome food source. After all, to fly at 50 to 70 wingbeats a second, they have to have food all day long.

Some species build up reserves in their crops to last the night; others go into a state of torpor to cut down their energy expenditure. These adaptations allow them to continue flying the next morning.

A study conducted in Brazil in the late 1980s established that a 1/3 ounce (10 g) hummingbird consumed 30 cubic inches (0.5 l) of oxygen per hour in its regular activities, or 10 times what a human consumes relative to body weight. To circulate this oxygen in the blood, a hummingbird's heart can beat up to 1,440 times a minute (compared with 150 for an exercising human). This vigorous activity maintains the bird's body temperature at 104°F (40°C).

Hummingbirds consume the most energy.

Another hummingbird species, the Frilled Coquette, is unable to build up enough energy to keep its body temperature up during the night. Instead, it halves its metabolic rate by going into a state of torpor, where it consumes 50 to 100 times less oxygen. However, returning to normal activity is laborious. With its reduced metabolic reserves, at first the hummingbird is unable to fly. To come out of its lethargic state and regain its normal body temperature, it must undergo a 15-to-20-minute period of intensive food consumption. And with its small reserves, it gets only one chance to start its engines. If it cannot do so, it dies.

ON THE MENU: BEESWAX, WORMS AND TADPOLES

What each bird species eats is normally a function of where it lives. However, taken as a group, birds feed on an enormous variety of organisms. Many species eat insects, but seabirds prefer the delicacies of the ocean: jellyfish, starfish, sea urchins, worms, mussels, fish and crustaceans. Frigatebirds are known for stealing fish from other fishing birds.

Some birds eat only seeds or fruit; many eat both. The New World warblers eat insects, but many will feed on berries after nesting and while wintering in the southern United States and Mexico. Raptors and shrikes are carnivores; their diets include other birds, frogs, tadpoles, lizards and snakes. Hummingbirds and other birds whose daily diet consists mainly of flower nectar will occasionally add insects to their diets. And then there are the scavengers—nature's garbage collectors.

The African honeyguides, small dark-colored birds, favor beeswax and bee larvae. Mammals and even humans will "follow the nose" of the Greater Honeyguide in order to locate beehives. Occasionally, the species will defend its treasure against other hungry birds.

THE BABY SNATCHERS

Widespread in North America, Asia and Europe (mainly Scandinavia), Golden Eagles prefer large prey and will even

The Greater Honeyguide feeds on beeswax.

take a fawn. These great raptors were once fabled to snatch and devour newborn babies.

The Guadelupe Caracara, a large raptor of the Caribbean, became extinct due to the belief that it posed a threat to humans. This was pure fantasy, say researchers.

Nevertheless, a Golden Eagle weighing 11 to 13 pounds (5 to 6 kg) is able to attack much larger quadrupeds. A study published in late 1993 cites the case of a Golden Eagle that killed a young caribou in the Parc de la Gaspésie, in Quebec, Canada, a park with the only remaining caribou population south of the Saint Lawrence River.

According to Michel Crête, a biologist with the Quebec Department of the Environment and Wildlife, who collaborated on the study, scientists already knew that coyotes preyed on young caribou. But certain facts suggested that Golden Eagles may do likewise.

In May 1989, a department technician in a helicopter spotted a Golden Eagle feeding on the carcass of a young caribou. Examination of the carcass showed that the caribou had died recently. Its thorax had been pierced, and its heart, liver and lungs partially eaten. When the technician returned a day later, the carcass had disappeared—probably carried off by the eagle.

Although the eagle seems to have been the culprit on at least this occasion, a second study by the department in response to low caribou survival rates suggested that coyote predation is largely responsible for endangering the herd of 250 animals.

In northern Scandinavia, however, the ecological relationship is different. There, the eagle is considered a major predator of caribou. According to studies conducted in Finland from 1970 to 1985, the 11 percent mortality rate among caribou, especially young animals, was due to predation by eagles and red foxes.

According to Michel Crête, newborn caribou are sometimes quite scrawny, weighing nine pounds (4 kg) or less, and would no doubt be easy prey for the eagles. He cites a Norwegian study in which a newborn caribou was found with its skull pierced, apparently by the talons of an eagle. Another caribou killed in the same way weighed 77 pounds (35 kg).

In the Yukon, Golden Eagles have been seen on three occasions killing young Dall sheep weighing about nine pounds (4 kg), and an eagle was seen taking off from a cliff with an animal carcass of about 22 pounds (10 kg) in its talons.

TURTLES TO HEDGEHOGS

The Golden Eagle is also known to attack white-tailed deer, pronghorn antelope, mountain goats and bighorn sheep. When live food is scarce, it scavenges carcasses and is sometimes killed in animal traps.

Farmers have unjustly accused it of countless crimes. American ranchers waged a merciless war against the Golden Eagle in the 1940s and 1950s because of its undeserved reputation as a predator of sheep. More than 20,000 eagles were killed from planes and helicopters.

A Montana study conducted in an area where 35,000 sheep were grazing failed to produce any evidence that eagles had killed a single animal. Instead, it was found that 70 percent of the eagle's diet consisted of cottontail rabbits.

The Golden Eagle is a solitary breeder and generally a silent bird. At 3 feet (1 m) long and with a wingspan exceeding 6 feet (2 m), it is one of the world's largest eagles. Although it usually soars on air currents, reportedly it can stoop at speeds of 199 mph (320 km/h).

The average clutch size is two eggs, and the age of first flight is 65 days. The nestlings are fed mostly small birds, but the diet of adults can be quite varied and includes dogs, cats, raccoons, rats, skunks, snakes, insects, even turtles. Nevertheless, rabbits and hares are the Golden Eagle's preferred food.

In Europe, Golden Eagles have a marked preference for marmots, young chamois, deer fawns, rabbits, fox cubs and hedgehogs. In some areas, turtles make up 20 percent of the summer diet.

In North America, Golden Eagles take a variety of avian prey, including Great Blue Herons, Great Horned Owls, Wild Turkeys, American Crows, geese, ducks, grouse and pigeons.

The Golden Eagle is an adaptable bird and can even hunt in groups if this serves its needs. Two eagles, for example, were chasing a rabbit, which attempted to hide in a small grove. One of the eagles perched on a branch while the other slowly approached the prey's hiding place. When the second bird flushed the rabbit out, the first one pounced on it.

RAVENS ON THE GREEN

Birds must often use considerable ingenuity to obtain food. Researchers cite the case of two hungry ravens watching a dog that was gnawing on a bone. One of the birds approached the animal and bit its tail. When the dog dropped the bone in pain, the other bird took off with the prize.

Remarkable as it may seem, some birds have even learned to use tools to capture prey. The most famous case is that of the Woodpecker Finch of the Galápagos Islands, a sparrow-sized bird that uses a cactus needle to dislodge its prey. The Egyptian Vulture of Africa and southern Europe breaks ostrich eggs by dropping stones on them.

In North America, Green Herons sometimes fish by tossing bait onto the surface of the water. Small fish that come to investigate are immediately captured. The Brown-headed Nuthatch will use a stiff piece of bark as a lever to flake off other pieces of bark in its search for insects. A tame Blue Jay used pieces of paper to pick up food left just outside the bars of its cage.

The Herring Gull has its own method of getting at hard-to-reach prey. It will drop mussels or sea urchins from a height onto rocky ground in order to break them. The Laughing Kookaburra, a 16-inch (40-cm)-long Australian species, kills snakes in a similar fashion.

This tactic is also used by ravens, often to the displeasure of golfers. On golf courses all over North America, players have noticed balls disappear, stolen by a raven or a crow. A raven

The Egyptian Vulture breaks ostrich eggs by dropping stones on them.

THE SECRET LIVES OF BIRDS

would pick up a ball, fly to a height of several hundred feet or more, then drop it. Then the bird would fly down to examine its booty, trying in vain to break the ball with its bill as if it were an egg.

On one course, ravens stole 8,000 balls in a single year.

Golf balls are often spirited away by ravens.

An Eye for an Eye

The penetrating gaze of raptors has fascinated us since time immemorial. Their ability to spot prey at great distances has been put to use for centuries in falconry. In many countries, raptors are still trained for that purpose.

The fierce gaze has also been a symbol of power in many civilizations. The Romans made the Golden Eagle their emblem; the United States chose the Bald Eagle, and Mexico the Crested Caracara.

Although there has been a tendency in the past to exaggerate the visual acuity of eagles and falcons, nevertheless it is at least three times greater than that of humans. They can also distinguish a wide range of colors.

But they are not alone among birds. Small songbirds, which must capture insects in order to survive, have equally acute vision. Hunting insects on the wing requires exceptional visual coordination.

Humans have about 200,000 cones/mm² on the wall of the retina. Cones are small light- and color-sensitive nerve cells that allow us to distinguish the details of an image. In contrast, House Sparrows have 400,000 cones/mm², and many raptors have up to one million. Game birds and seed eaters, however, are less well endowed. Rock Doves, for example, have poorer vision than humans and their eyes take much longer to adjust to the dark.

People can see clearly only those objects located in a small part of their field of vision and must move their eyes constantly to get an overall picture. Birds, however, can take in at a glance everything in a 20-degree sweep, or eight times more than what humans can. In this way, they get a much more panoramic view than we do and can perceive movement with much greater ease, without having to move their eyes—an essential attribute for a hunter.

THE FARSIGHTED SHRIKE

The Northern Shrike, whose visual acuity is as great as an eagle's and a falcon's, can spot a flying bumblebee at a distance of 450 feet (140 m). One was seen trying to capture a caged mouse it had located at 900 feet (275 m) away from its perch. Apparently, shrikes will react to other shrikes 1,900 feet (570 m) away. One wild shrike reportedly flew almost a mile (1.3 km) to attack another one in a cage.

Birds' retinas also contain rods, a different type of nerve cell that focuses light and

The Great Horned Owl has better eyesight by day than by night.

enables birds such as owls to perceive poorly lit objects. At night, Great Horned Owls have three times better vision than humans, although they can see nothing at all in total darkness. But without their extremely well-developed sense of hearing, nocturnal raptors would probably have great difficulty finding enough food. And despite their finely honed senses, they usually come up with a catch only once every five tries.

Contrary to popular belief, owls have enough cones in their retinas to see very well during the day. In fact, Great Horned Owls see better by day than by night. A Great Horned Owl, for example, was found to detect a hawk in broad daylight that was flying so high a human observer could not see it.

ONE EYE OR TWO?

Sight is the most important sense in birds, even nocturnal raptors. It is crucial in finding food and mates, maneuvering in flight, detecting predators and hunting. Appropriately, the eye takes up a large part of a bird's skull. The eyes of a European Starling, for example, account for 15 percent of the weight of its head. In certain species, the two eyes weigh as much as the entire brain and can be as big as human eyes.

Birds are equipped with both monocular vision—in which the eyes see and move independently of each other, always horizontally—and binocular vision straight ahead. American Robins on a suburban lawn trying to make out a nearby object will use each eye in turn, rapidly turning their heads. In many owls, the eyes cannot move.

Visual fields vary, depending on the species of bird. The eyes of an American Woodcock are located at the back of its head, providing it with a nearly 360-degree view of its surroundings. In contrast, Great Horned Owls have poor monocular vision, since their eyes point straight ahead. Owls makes up for this by being able to rotate their heads 270 degrees.

Owls have exceptional binocular vision, however, due to the position of their eyes. When both eyes can see the same object at the same time, as in humans, the depth of field increases considerably. This allows the viewer to distinguish the contours of objects and gauge distances more accurately. In owls, 60 to 70 percent of their vision is binocular. In raptors and certain large insectivores it is 25 to 50 percent, and in seed eaters it is 10 to 30 percent.

A DIVING MASK

The eyes have to be constantly cleaned and moistened. In birds, this function is performed by the *nictitating membrane*, a third eyelid folded away in the corner of the eye nearest the bill. It is transparent in diurnal birds. This membrane constantly sweeps the eye horizontally or diagonally, cleaning and moistening the cornea without overly interfering with sight. If a bird's head is moving, the membrane moves more often.

The nictitating membrane also plays a protective role. In the Common Wood-Pigeon, it slides across the cornea at the very moment the bird reaches down to pick up food.

The Loggerhead Shrike's visual activity is truly exceptional.

Like many other water birds, the Razorbill, a widespread species of the North Atlantic and the Saint Lawrence estuary, has a membrane with a tiny central window that allows it to see properly underwater—a little like a diving mask.

When birds sleep, their two vertical eyelids are shut. In diurnal species, the lower eyelid moves to meet the upper; in nocturnal species, it is the other way around.

The Razorbill has its own "diving mask."

No Vocal Chords, But Two Different Voices

Most birds sing, and some are even virtuosos. Some species are silent, however. Wood Storks and carrion-eaters such as Turkey Vultures, Lammergeiers, Egyptian Vultures and Cinereous Vultures are almost entirely silent, apart from a few croaks.

Large pelagic birds like fulmars and petrels, which have external nostrils to expel salt, remain silent for most of their lives. The Razorbill is another member of this largely silent minority, although it occasionally emits a few grunts.

Unlike humans, birds do not have vocal chords. The way they produce sounds is based on a peculiar anatomical phenomenon. In fact, many birds can sing with their bills closed or even full of food.

WHERE IS THAT VOICE COMING FROM?

The human voice is produced by the vibration of the vocal chords in the larynx, located in the upper part of the windpipe, or trachea.

Birds, on the other hand, produce sounds through ventriloquism. Their sound box is an enlargement of the trachea at the point where the windpipe divides into two bronchi. This organ is called the *syrinx* and is exclusive to birds. It is a cartilaginous chamber with elastic membranes. It is controlled by muscles that vary the tension and position of the tissues that vibrate with the passage of air.

The syrinx is composed of two independent parts and can produce two distinct, yet similar, songs simultaneously. The two sources of sound can also fuse to produce a single song, as was discovered only a few years ago.

As Frank B. Gill notes, this phenomenon was first observed in Brown Thrashers and was later found to occur in grebes, bitterns, ducks, sandpipers and some songbirds. It also occurs in the neotropical bellbirds, whose loud, bell-like song can be quite startling.

Generally, birds sing by expelling air from their lungs. But unlike us, their nasal cavities, bill and throat are used to modify, rather than produce, sound. The quality and complexity of a species' song usually increase with the number of syrinx muscles the species has. Pigeons, for example, have only a single pair of these muscles, while most passerines, including gifted vocalists such as the American Crow and Gray Catbird, have from five to nine pairs. The European Starling, a master of mimicry, has seven to nine pairs of muscles.

The silent Turkey Vulture has no syrinx at all, and storks and ratites are entirely lacking in syrinx muscles.

Vocal volume is more a result of resonance than the pressure of the air leaving the lungs.

BATTLE CRY

What is the difference between a call and a song? As ornithologists define it, a call is brief (rarely more than five notes) and conveys a message. Calls may be used to intimidate aggressors, warn young birds to hide in the event of danger, identify food sources or keep family flocks together, as with the domestic chicken. In gregarious species or in migratory flocks (geese, for instance), calls help birds keep one another in sight.

The ability to call and understand calls is innate. European Starlings, grackles and cowbirds react instantly to the distress calls of other birds, whether of their own species or another.

Though the object of many a poet's rhapsody, bird songs often have little to do with courting. Quite the contrary. They are usually cries of war.

A male secondary sex characteristic not usually shared by the females, song is controlled by certain hormones. Its primary purpose is to inform other males of the species that the singer controls a certain territory with invisible boundaries—a musical No Trespassing sign, so to speak.

Song also serves to attract females for mating purposes. It helps to maintain the pair bond during nesting, as well as the social cohesion of the species.

The Northern Cardinal is an accomplished singer.

Hum a Few Bars

Although a species' musical repertoire is partially innate, it is often enriched by original compositions. Not only is song an individual characteristic, but many species have regional dialects. Song can be used to identify parents and chicks, or members of a pair and their neighbors. Studies have shown that, in many cases, young birds isolated from their parents at an early age never learn to sing correctly.

The repertoires of birds are more complex than they may appear. An individual Song Sparrow, a familiar North American species, produces eight to 10—sometimes up to 20—different melodies. More than 200 versions of the Song Sparrow's song have been catalogued. Furthermore, individual Song Sparrows can detect and differentiate their neighbors' songs to avoid territorial disputes.

Wowing the Women

Jim Mountjoy, who has studied thousands of recordings of starling songs for years as part of his doctoral thesis at McGill University in Montreal, Canada, has found that older, more experienced males have wider and more complex repertoires. These older males attract more females than do their less experienced competitors.

Other males recognize the biological superiority of these avian Carusos and leave them alone. Why pick a fight if you know you are bound to lose?

Female starlings also benefit from mating with these virtuosos. By choosing a better singer as a mate, they increase their chances of locating a safe and suitable nest site.

Older, more experienced males have other advantages. They tend to winter where they nest, whereas first-year birds migrate. When the young birds return in spring, the older males have already staked out a territory and the young males must battle for a spot.

Older males often defend several nest sites and have even been known to mate with more than one female, but these bigamous birds never exceed 5 percent of the population.

The song of dominant males also attracts younger males, which follow their elders to discover the best nest sites. Since the older males are soon busy raising their own families, they eventually relinquish less suitable nest holes to their rivals.

Sing a Long Song

Bird song is most apparent in spring and early summer, when territories are being established, pairs are formed and nesting occurs. It tapers off to silence as the molt approaches. There are exceptions, however: gregarious and itinerant species sing all year long, even in the dead of winter. Still, song activity is less intense in winter than during the breeding season.

A song can be defined as a fairly elaborate sequence of notes, often grouped into musical phrases repeated at regular intervals. Most songs last about four seconds, but the Winter Wren's can last up to 10 seconds.

Arthur Cleveland Bent, the author of a series of books on North American birds, reported that a Rose-breasted Grosbeak's song is 2 to 6.8 seconds long, while a Northern Cardinal's song lasts from 1.8 to 4.2 seconds.

The same song is often repeated many times. One Song Sparrow sang the same song 2,305 times on one May day, and observers counted 3,000 repetitions of one warbler's song in 16 hours. But the record is held by a Red-eyed Vireo that sang 22,197 times in a single day.

Birds will sometimes produce a sweeter, more elaborate, version of their primary song during bad weather or excessive heat. They normally begin to sing before sunrise, and the bird song gradually falls off until noon. It picks up later and continues until the end of the day.

Males in their breeding territories will sing at almost any time if an intruder appears. Although most species usually sing from a perch, some species (American Robin, Blackbird, Vesper Sparrow) will do so from the ground.

Some diurnal birds such as the Northern Mockingbird also sing at night. It is not uncommon to hear forest-dwelling White-throated Sparrows singing in the middle of the night in a big city during spring migration. Perhaps the birds mistake the lights of the city for broad daylight!

DUETS

About 200 of the world's bird species sing duets, including the House Sparrow. Gonoleks, African birds similar to our shrikes, perform exceptional duets in which the male and female take turns singing portions of the same song in such a

The song of the Rose-breasted Grosbeak lasts anywhere from 2 to 6.8 seconds.

coordinated fashion that it sounds as if only one bird is singing. The pauses between phrases are so short that they are almost impossible to detect.

Who is the best singer? The answer is very subjective, and many contenders for the title are to be found on every continent. According to one American researcher who studied and classified the songs of 70 North American species, the three greatest virtuosos are Bewick's Wren, a small bird found from southern Ontario to Mexico; the Winter Wren, a widespread species of Europe and North America; and the Northern Mockingbird, a master of mimicry.

STOP REPEATING EVERYTHING I SAY!

A number of bird species are excellent imitators. In Europe, the Marsh Warbler is undoubtedly number one, since it is able to reproduce the songs and calls of 100 bird species of Africa, where it winters. Familiar species imitated include the Blackbird, Barn Swallow and Blue Tit.

The European Starling is another skilled mimic. It can imitate not only the songs of 50 bird species, but also a cat's meow, a dog's bark and a cow's moo. Starlings in captivity can even be trained to say a few words.

Starlings have been heard to mimic the drumming of a woodpecker by tapping on metal siding—even though their bills are poorly designed for this tiring activity. Starlings will imitate everything from House Sparrows to Herring Gulls, including such familiar North American species as Killdeer, Red-winged Blackbird and American Goldfinch.

MOZART'S STARLING

In England, a pet European Starling not only learned a few words, but could even recite a rhyme or two. Its imitation of a ringing telephone was so convincing that its master inevitably ran to pick up the receiver.

The most famous starling is probably the one that belonged to the great composer Wolfgang Amadeus Mozart. Mozart bought the bird on May 27, 1784, in a pet shop. As the story goes, the composer was charmed by the bird's ability to imitate parts of his Piano Concerto No. 17 in G Major. No one knows how the bird managed to learn it, but it may have heard Mozart whistle or hum the score, as was his habit, on a previous visit to the pet shop.

The influence the bird had on the composer's music is unknown, but Mozart appears to have been very attached to his pet. When the starling died, it was given a ceremonial burial and Mozart even dedicated a poem to it.

The Northern Mockingbird is surely the most accomplished mimic in the bird world—so accomplished, in fact, that its songs cannot be distinguished from the real thing, even by electronic analysis.

Widespread throughout North America south to Mexico and the Caribbean, the Northern Mockingbird has been pushing its breeding range northward in the past few years and now nests in southern Quebec.

As its name implies, the mockingbird mimics dozens of bird species (some suggest up to 150) and can croak like a frog, chirp like a cricket, bark, meow, cackle like a hen and imitate squeaky wheels and piano notes.

Even if you listen carefully, you cannot always tell which species is being imitated, since mockingbirds often mimic only a small part of each song. Researchers have heard one mockingbird reproduce the call or song of 23 species in 10 minutes. Its repertoire changes and increases with age.

UNKNOWN ARTISTS

Aside from these three outstanding mimics, there are hundreds of other species less skilled in the art, and it is estimated that 15 to 20 percent of the world's passerines practice some form of vocal mimicry.

In North America, the Gray Catbird is particularly talented. Curiously, its own call, which sounds like a meowing cat, is not an imitation. Catbirds can copy the songs of 40 species, including the Blue Jay, Barn Swallow, American Robin, even frogs.

The Brown Thrasher is more limited in its abilities. It is far surpassed by the tiny Carolina Wren, whose repertoire includes the American Kingfisher, Red-winged Blackbird, Baltimore Oriole, Eastern Bluebird and even, in a curious turning of tables, the Gray Catbird!

Like some of its Old World congeners, the Loggerhead Shrike can also mimic the sounds of other birds, including the Eastern Kingbird, Red-eyed Vireo and Eastern Bluebird.

Corvids are also good mimics. Blue Jays have reportedly mimicked the calls of Red-tailed Hawks, Whip-poor-wills and American Robins. American Crows can imitate a dog's howl, a baby's cry or a hen's cackle.

In Europe, the Eurasian Jay, Sedge Warbler, Common

Redstart and Red-backed Shrike are also considered excellent mimics.

Terres reports that a European researcher was intrigued by Blackbirds uttering a strange whistle. After inquiring in the neighborhood, he found the source—a man who for years had been whistling for his cat.

There is also the case of the Eurasian Jays studied by the famous ethologist Niko Tinbergen and his students at a camp in the Netherlands. Tinbergen was in the habit of waking up his students by whistling and tapping on the roofs of their tents. One morning at four o'clock, one jay imitated this habit of the professor's so adeptly that the students came running out of their tents.

Ornithologists still do not know the reasons for bird mimicry. What is certain is that birds must hear a song at least a few times before they can imitate it. One theory about the Northern Mockingbird is that its imitations may help singers distinguish one another, which could play a role in the species' social organization.

It has been suggested that Marsh Warblers imitate African birds to locate males that winter at the same location.

TALKING BIRDS

A number of birds can imitate the human voice. Although parrots have long been recognized as skilled imitators of words, they are not the only ones that can talk.

Some budgies can build impressive vocabularies.

The Hill Myna, a member of the starling family native to Asia, has become a popular cage bird the world over because of its ability to speak. One mynah was famous for being able to sing an entire song.

Like European Starlings, ravens and crows trained at an early age can learn to imitate a few words, and even mimic laughter.

But the hands-down champions are still the parakeets and parrots—especially those with a patient teacher. Certain birds are undeniably more talented than others. For example, one parrot, after only a few days of instruction, could repeat whole sentences, and even count to 10. Another very gifted subject could pronounce more than 500 words clearly, as well as recite rhymes.

STOP ACTING LIKE A BABY

Research shows that Budgerigars in Australia imitate one another in their natural environment. This practice appears to strengthen pair or group bonds in these gregarious birds. However, pet parakeets and parrots have more developed vocabularies because of their close relationship with humans.

Birds seem to understand that by repeating words, they will attract their owners' attention. In fact, parakeets are apparently more talkative when their owners leave the room, no doubt trying to make them return.

This behavior is similar to that of newborn babies. A baby's first efforts at speech serve to attract its mother's attention and care. It is said that the best way to get your pet bird to talk is to feed it by hand while saying the words you want it to learn.

According to recent research, some birds have remarkable

powers of comprehension, and in some circumstances can even associate ideas. This certainly suggests that birds have some level of intelligence. Here are two examples, one related by a French-Canadian woman whose budgie demonstrated amazing talents.

LOULOU AND ALEX

Several years ago, Lucille Hétu took advantage of being on sick leave to teach Loulou, her young male Budgerigar, to talk. After a few months of training, Loulou began to speak his first words, and after three years of intensive schooling, he had a vocabulary of 130 words, could recite 38 sentences and sometimes spoke for 25 minutes nonstop.

One evening, when his owner was nodding off while reading her newspaper, Loulou said, "Go to sleep." Other times, while watching his mistress cook, the budgie would ask, "What are you doing?" Some mornings, on leaving his cage, Loulou would perch on Lucille's shoulder and exclaim, "Good morning!" Then he would ask, "Did you sleep well?"

Loulou also used to repeat "He's gone" whenever Mrs. Hétu's husband left the house. One day after he had stormed out in a bad mood, Loulou explained, "He's gone mad."

Alex the Gray Parrot is known throughout the scientific world for his feats of intelligence. Trained for several years by Irene Pepperberg, a University of Arizona researcher, Alex has learned to do something that seemed unattainable by talking birds—associate ideas. In many respects, Alex has surpassed monkeys in his demonstrations of intelligence.

For instance, he has learned to say "No" when he does not

want to be touched. He can identify and ask for more than 80 objects of different colors, shapes and materials. When asked the color of a square object among other objects of different shapes and colors, he usually answers correctly.

Alex can even tell if shapes are similar or not. In 1994, he was taught to count by associating each digit with a specific number of objects. Thus he can establish a relationship between the digit, the group and the color of the objects represented by it.

Alex, a Gray Parrot, is capable of remarkably elaborate speech.

THE SECRET LIVES OF BIRDS

Speak Up— I Can't Hear You

Hearing is much less acute in birds than in mammals in general and humans in particular, although owls are a notable exception. Humans can perceive frequencies ranging from 20 to 20,000 Hz. Most birds cannot even come close to this mark.

European Starlings and House Sparrows hear about the same range of high frequencies as we do, but their ability to perceive sound decreases considerably at low frequencies, and they cannot hear sounds below 650 Hz. Rock Doves and domestic chickens, however, are able to hear frequencies below the range of human perception.

The Hairy Woodpecker and other woodpeckers can hear very low frequencies, and are believed to detect wood-boring insects by sound. Such insect sounds are inaudible to the naked ear but can very easily be heard through a stethoscope.

Unlike bats, birds do not perceive ultrahigh frequencies (above the range of human hearing). They are, however, sensitive

THE SECRET LIVES OF BIRDS

to changes in pitch, although here, too, human hearing is superior. On the other hand, birds find it much easier than we do to locate the source of a sound.

THE BETTER TO HEAR YOU WITH, MY DEAR

Although auditory capability varies from species to species, hearing is, in general, a bird's second most important sense, after sight.

Birds' ears do not protrude externally the way ours do, and except in ostriches and Old World vultures, they are covered by feathers. However, these feathers have no barbules, which would otherwise obstruct sound. In certain diving birds such as penguins, the circular muscles of the external ear completely close the orifice when the bird is underwater.

The avian ear also has a membrane similar to the eardrum, as well as semicircular canals, which play a vital role in balance. There appears to be a relationship between the size of these canals and aerial prowess: for example, the canals are larger in falcons than in ducks.

Unlike other birds, nocturnal raptors, especially Great Horned Owls and Barn Owls, have a very well-developed sense of hearing. This allows them to capture prey in total darkness.

Like its facial disk, the Barn Owl's ears are asymmetrical, with the left one slightly higher than the right. Barn Owls are more

The Barn Owl's left ear is set higher than its right one.

sensitive to certain noises than other owls and can easily detect high-frequency sounds. Great Horned Owls and Eurasian Eagle-Owls are very sensitive to low frequencies—even lower than those humans can detect.

In 1992, a blind, starving Great Horned Owl attacked an 11-year-old schoolgirl in broad daylight in a village in Canada. The bird's eyes had been pierced by porcupine quills and it located the girl by her footsteps only, believing her to be prey. Obviously not in its normal state, the animal had to be killed. The girl escaped with no more than a scar on one ear.

ECHOLOCATION

Oilbirds and certain swifts are unique in the bird world. They live in caves and, like bats, they navigate using echolocation, which involves locating obstacles by emitting ultrasounds that produce an echo. The swifts in question live on the coast of Malaysia, and their nests are considered a delicacy among Asians.

Oilbirds, which look like large nighthawks, are about 12 inches (30 cm) long and have a wingspan of about three feet (1 m). They inhabit South America and a few Caribbean islands.

In Venezuela, one of the caves is open for

Oilbirds use a method of orientation similar to the sonar of bats.

tourists, and the show the birds put on leaves a lasting impression on visitors. They witness the fascinating spectacle of the birds calling as they fly around the ceiling in total darkness. Every few seconds, each bird emits a sound that echoes off the cave walls, giving the bird feedback about its location.

This method of navigation is, however, much more poorly developed than in bats; although oilbirds can detect larger objects (over 3/4 inch [2 cm] in diameter), they collide with smaller ones.

Oilbirds leave their caves at nightfall in search of food—fruit, in particular—which they bring back home to feed on. Unlike bats, they do not use radar to locate their food, but instead are thought to use their sense of smell.

Smell:
The Overlooked Sense

S ense of smell in birds has not attracted the attention of many ornithological researchers. For a long time it was thought that most birds were incapable of perceiving smells. The most interesting discoveries in this area have been made only in the past decade.

It has long been known, however, that many pelagic species, such as the albatrosses and the Atlantic Puffin, are able to smell, and that the Brown Kiwi of New Zealand, a nocturnal bird, uses its sense of smell to locate prey. Unable to fly, kiwis use the nostrils at the end of their bills to detect smells—an exception in the bird world. They feed mainly on earthworms, which they locate by their smell. Moreover, kiwis (of which there are three species) can detect strong-smelling bait placed several inches underground, but ignore odorless bait.

Until recently, it was also believed that passerines such as sparrows and Old World warblers were unable to perceive smells.

The justification for this view was that the olfactory lobes of their brains are very small compared with those of mammals and other birds known to have a sense of smell.

FRAGRANT HERBS

Recent data suggest otherwise, however, and several experiments have shown that, despite their tiny olfactory lobes, many passerines can detect certain smells as well as rabbits and rats can. In fact, researchers now believe that most birds perceive smells and use this ability in their daily lives, even if sense of smell is less important than sight and hearing.

European Starlings, for example, choose certain nesting materials with the help of their sense of smell, sometimes using fragrant plants that inhibit bacterial growth or prevent the eggs of parasitic mites from hatching. Mites can cause nestlings to lose up to 20 percent of their blood.

Black-billed Magpies can smell rotten meat.

Although it was always suspected that certain carrion eaters use their sense of smell to find food, there was little concrete evidence. In the mid-1980s, however, a Turkey Vulture, a widespread North American species, was used to detect a leak in a gas pipeline. A product with the odor of rotting food was put in the pipeline. Forty miles (65 km) away, a group of vultures began to circle above the leak. Magpies can also sniff out the odor of rotting meat.

ORIENTATION BY SMELL

Laboratory tests with Canaries, Turkey Vultures, domestic chickens, pigeons and ducks have shown that these birds detect smells. Sexual excitement in Mallards was found to be largely related to a scent given off by females during the breeding season.

Pelagic birds such as fulmars and albatrosses can easily locate fish oil on the ocean surface. Leach's Storm-Petrel, a marine species about 8 inches (20 cm) long, can detect a raft of tiny crustaceans at distances up to 15 miles (25 km) away. Although the crustaceans live underwater, the smells they emit are apparently easily detected by birds once the smells are dispersed in the air.

It is believed that these storm-petrels, like the Atlantic Puffin, locate their burrows by smell. In one experiment, storm-petrels whose olfactory nerves had been temporarily blocked were unable to find their nests, whereas the control birds had no problem doing so.

There is a great deal of evidence to suggest that other species can perceive smells well. African Honeyguides, which feed on beeswax, have been attracted many times by the candles lit by missionaries saying mass on the African savanna.

TASTY?

We have little information on sense of taste in birds and the role it plays.

The Blue Jay, a familiar species found throughout eastern North America, always reacts with disgust when it captures a Monarch Butterfly for the first time. The insect is immediately spit out and the bird never tries to eat one again.

Several species, including chickens, pigeons, the Great Tit and quails, reject bitter food. Other birds, such as hummingbirds, prefer sweet substances, while some consume food with varying degrees of acidity.

Ornithologists tell us that birds are much more poorly equipped for tasting than mammals. Humans have 9,000 to 10,000 taste buds; rabbits and rats have 17,000 and 1,200, respectively. In contrast, Mallards have 400 taste buds, or a little more than hamsters. The Japanese Quail has 60, pigeons fewer than 40 and domestic chickens only 24.

Quail have very few taste buds.

Waiting to Exhale

Like humans, birds breathe through their nostrils and absorb oxygen through their lungs. But in other ways their respiratory system is very different from that of mammals.

Hummingbirds breathe 143 times per minute; turkeys do so seven times per minute; the figures for ostriches and humans are six and 12, respectively.

Because birds expend so much energy, their oxygen needs are quite high. The air sacs are organs unique to birds. They are essential in respiration, even though they do not take part in oxygen/carbon dioxide exchanges, which is the work of the lungs. Composed of a thin tissue of cells, these air sacs, of which there are usually nine, make birds lighter.

Because of their high cell density, the lungs of birds take up half as much space as humans' do relative to body size.

Whereas humans exhale a large part of each breath, birds completely expel the air from their lungs. Not only does this air

circulate one way only, but it also takes two inhalations and two exhalations to complete the cycle. On inhalation, fresh air enters the air sacs; on exhalation it goes into the lungs. At the same time, stale air is pushed out by the lungs and collected in other sacs, to be expelled.

In this way, a bird's body absorbs maximum oxygen. This is why birds can breathe without difficulty at high altitudes, where the oxygen content of the air is low.

The air sacs branch into the bones of the wings and legs, and play a vital role in temporarily storing air. Their position within the body also helps birds maintain their heat balance, particularly during flight—a function as important as breathing itself.

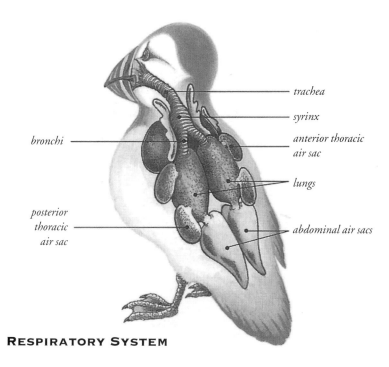

bronchi

posterior thoracic air sac

trachea

syrinx

anterior thoracic air sac

lungs

abdominal air sacs

RESPIRATORY SYSTEM

The air circulating through the body recovers and dissipates heat. Without this function, birds would succumb to the intense heat they generate in flight. A passerine produces nine or, in some cases, even 15 times as much heat in flight as at rest.

STOUTHEARTED BIRDS

Relative to their body weight, birds' hearts are much more powerful than those of humans because their respiratory requirements are greater. For this reason, a bird's heart is much bigger—often twice as big—as that of a mammal of comparable size. The human heart normally beats 72 times per minute. In contrast, the average for birds is 220. Representative figures are

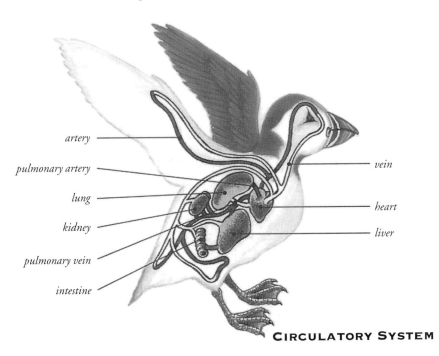

artery
pulmonary artery
lung
kidney
pulmonary vein
intestine

vein
heart
liver

CIRCULATORY SYSTEM

700 beats per minute in the House Wren, 220 in the Rock Dove and 514 in the Canary.

A hummingbird's heart beats 615 times per minute on average, but in smaller species, it can reach 1,400 beats per minute during intense activity. The heart of a Cuban Bee Hummingbird—the world's smallest bird, weighing just under one ounce (2 g) and measuring two-plus inches (5 to 6 cm)—beats at least 10 times faster than a human's. In ostriches, which can weigh up to 330 pounds (150 kg), the heart rate is only 35 beats per minute.

Blood pressure in birds is high because of the speed of their hearts. Heart attacks can result from fright or intense excitement. One warbler died of a heart attack during the breeding season and "overweight" turkeys sometimes meet the same fate.

PANTING

Air sacs help control a bird's internal temperature, but the animal's body must also constantly adjust to the outside temperature. With such high metabolisms, birds must expend a lot of energy on maintaining homeostasis. This is problematic in times of extreme cold or heat. Birds have no sweat glands, so their respiratory system is responsible for heat regulation.

The body temperature of humans is 98.6°F (37°C). In birds it is usually around 104°F (40°C). In small species, it often reaches 107°F (42°C), but at night drops to 102°F (39°C).

Panting, a phenomenon easily observable in familiar birds, helps them tolerate excessive heat. I recall a Barn Swallow that had made its nest on a beam inside a barn, a few inches from the metal roof. On hot days it would perch, immobile, inside the broiling barn, its bill open. The excessive heat acted as an incubator.

Since panting expels moisture from the lungs, birds must obtain sufficient water. Blood circulation to their feet sometimes increases to help dissipate heat.

Species such as owls, the Great Blue Heron, Cattle Egret, pigeons and doves have a remarkable method of dissipating heat. They expel, via their mouths, the layer of skin covering the inside of their throats to increase the surface area of their skin that is in contact with the air.

UROHYDROSIS

During heat waves, birds tend to hold their feathers tight against their bodies to make it easier for heat to dissipate through their skin. Cormorants, Turkey Vultures, African ibises and Marabou Storks spread their wings to bring their apteria into contact with the air. Other species turn their backs to the wind, allowing it to fluff up their feathers and refresh their skin.

Birds usually become less active when the temperature exceeds 91°F (33°C) and look for a cool place to rest. Semidesert-dwelling sparrows will even escape the heat inside a prairie dog burrow. Vultures take advantage of rising air currents to soar at higher, cooler altitudes.

The phenomenon of urohydrosis is observed in the Marabou Storks of Africa.

The situation is even more critical when air temperature exceeds body temperature. Then, in addition to sheltering in the shade, birds must find enough water to expel more heat through their lungs.

Certain waders, such as the American Stork as well as the Black and Turkey Vultures, use a peculiar method to cool themselves—they wet their legs with their own liquid droppings. This method, called *urohydrosis*, lowers the temperature of their legs by one degree, something that could not be achieved by panting alone.

Herons drop their excrement into their large nests. Researchers believe that the water vapor that is produced as the excrement dries refreshes the nest's inhabitants. Generally speaking, birds cannot survive when their body temperature exceeds 113°F (45°C).

DEADLY SLEET

Winter can be a difficult time. Birds in northern climates sometimes have to endure temperatures down to −4 to −20°F (−20 to −30°C) in January and February. Under such conditions, the fat reserves built up under their skin during the fall are not sufficient for survival.

It can be particularly challenging for animals to survive the cold on 14- to 15- hour-long winter nights. Although House Sparrows are more densely feathered in winter (about 3,300 instead of 3,000 during summer), they cannot survive longer than 15 hours without eating; at −20°F (−30°C), they can endure only seven hours.

At such times, House Sparrows must find a sheltered place to spend the night or they will die—something that happens regu-

The syrinx is the sound-producing organ in birds. It can simultaneously produce two separate, but very similar, songs, like in the American Bittern. (Page 70)

The European Starling is a renowned singer and an excellent mimic whose talents were even recognized by the great Wolfgang Amadeus Mozart. (Page 76)

The most brilliant mimic in the bird world is the Northern Mockingbird, which, some suggest, can imitate the songs of at least 150 species. (Page 76)

It was only recently proven that the Turkey Vulture, a carrion eater, can find its food by smell. (Page 88)

Birds' sense of taste is much more poorly developed than in mammals.
The Great Tit, however, will reject bitter foods. (Page 90)

Mating takes many forms in birds. This Common Tern is bringing a fish to its mate for an intimate dinner. (Page 104)

The Short-eared Owl (young pictured here) uses very little material to build its nest. (Page 107)

To gather enough mud to build their nests, Barn Swallows make up to 1,200 trips. (Page 108)

larly, especially to weaker or diseased birds. During one winter in the 1960s, many dead sparrows and starlings were found on mornings after especially cold nights. Interestingly, all the dead birds had been suffering from salmonellosis, a bacterial disease that had decreased their resistance to the cold.

Major snowstorms and frosts can also kill many birds if they are prevented from finding food for too long.

House Sparrows that winter in cold climates need three times more energy to survive the day than those living in the tropics.

GOOSE PIMPLES

Species that winter in Canada adapt to the cold in a number of ways. Resident American Goldfinches, for instance, build up more fat reserves under their skin than do migratory members of their species.

Under ordinary circumstances, a bird's body temperature usually drops during the night, often by two or three degrees, and this slowdown in the bird's metabolism helps it survive the winter. In certain birds, the drop in body temperature is even more pronounced in severe cold. To conserve heat better, birds tuck their heads under one wing and arrange their feet so that they are snugly covered by their feathers. The circulatory system of ducks and geese is designed so that their feet will always be sufficiently warm to resist freezing.

One of the first responses a bird has to the cold is to puff out its feathers, a reflex similar to the "goose pimples" (an apt phrase) that appear on human skin. This traps air on the surface of birds' skin and improves insulation. Chickadees avoid the wind by taking refuge among dense conifers or in natural cavities. Grouse, ptarmigan, Common Redpolls and Boreal Chickadees prefer to spend the night in the snow because of its insulating properties. In Alaska, the air temperature 23 inches (60 cm) under the snow was found to be –23°F (–5°C) when the ambient temperature was –58°F (–50°C).

Ruffed Grouse will dive into the snow to find shelter at the end of the day. If the snow is covered with a thick crust, the impact can be fatal. Sometimes grouse become trapped under the snow when such a crust forms during the night, and will die.

Ptarmigan often spend winter nights under the snow.

BIRD BUNDLES

Certain species, such as the Eastern Bluebird, spend the night in a shelter (sometimes a nest box), using one another's body heat to fend off the cold. One night, 100 Pygmy Nuthatches, a species of western North America, were found so packed together in the same refuge that some had died of asphyxiation.

One study in Russia showed that two Goldcrests huddling at 32°F (0°C) reduced heat loss by 23 percent; three cut the heat loss by 37 percent. Certain swallows will also huddle together when the temperature is too cold.

In one laboratory experiment with House Sparrows, 50 percent of the birds exhibited huddling behavior when the temperature fell below –14°F (–10°C); only 10 percent of them did so at 32°F (0°C).

HIBERNATION

Other species store food toward the end of the day as a way to maintain their temperature at night. Evening Grosbeaks store seeds in their crops. Common Redpolls, thought to be the most cold-hardy passerine, use their elastic esophagus.

One species stands out in its approach to the cold: the Common Poorwill appears to be the only bird that hibernates. This insect eater, weighing 2 ounces (60 g) and measuring 8 inches (20 cm), lowers its body temperature five or six degrees for two or three months while at its wintering grounds in southern Texas and northern Mexico. It can flee abruptly if disturbed, but requires several hours before its body temperature returns to normal.

Couple Life

The "love life" of birds is rather turbulent. It is punctuated by winged pursuit, intimidation, interminable song, dance, aerial acrobatics, even "intimate dinners." There are countless facets to the art of seduction.

Once a male has chosen a territory, he adopts a perch and begins singing to announce to others that he is the new owner. Those who challenge him are pursued. If they resist, there is a tussle. Male House Sparrows can actually come to blows. Although normally cautious, they can often be observed on the ground, locked in combat and oblivious to the presence of humans. The struggle lasts for some time before the birds realize their imprudence and fly away.

Rarely does the battle go further than a few taps of the bill, but every now and then the death of one adversary ensues. Usually, males chase off their rivals by singing or briefly pursuing them.

Occasionally, bellicose males have been observed attempting

to attack their own images in a pane of glass or a mirror. One springtime, a White-throated Sparrow was seen to repeatedly and resolutely attack an imaginary adversary—its own image—in the window of a car parked on a forest road. The Northern Cardinal, a highly territorial species, will behave in a similar fashion. And one owner of a country home with above-ground basement windows noticed that the screens had been torn by a crow that had tried to attack its reflection.

INSPIRING DANCES

To mark off their territory and attract females, woodpeckers sometimes tap noisily on a hollow tree trunk, the wall of a wooden house or an aluminum roof. The Ruffed Grouse drums by beating its wings back and forth. The sound this makes can easily be heard half a mile away in spring. In addition to producing sound in this way, the males show off their colors.

Pairs of other species such as loons, swans and certain grebes do wild dances on the water or perform aerial ballets, as do ravens and several raptors. Every breeding season, without fail, the Common Snipe rises into the sky, then plummets many feet while emitting a peculiar sound produced by the passage of air through its feathers.

The woodcock emits a nasal call before flying high in the air and then making an unusual whistling sound while spiraling downward. Male turkeys and peacocks parade before their mates with their magnificent feathers held erect.

In cranes and other large waders, the dance movements are complex and spectacular. American Indians, Australian aborigines and several African tribes have based some of their traditional dances on the breeding behavior of cranes.

In North America, the male Sharp-tailed and Sage Grouse (or in Europe, the Capercaillie and Black Grouse) perform a ritual in a clearly defined arena before a discreet gathering of females. The males parade around the center, trying to impress one another by raising their feathers or their bright-colored caruncles. They also show off the bright-colored bare patches on their neck.

The male Black Grouse performs spectacular courtship rituals during mating season.

The ritual, which takes place several times a day, alternates with dancing. The most dominant males often occupy the center of the arena. Eventually, a female enters to copulate with one of them. The female will lay her eggs several weeks later and raise the family.

This same ritual also occurs in the Ruff, a medium-sized shorebird with splendid spring colors.

DINNER BY CANDLELIGHT

Other birds have more intimate mating habits. In pigeons, doves, Northern Cardinal, crows, Common Redpoll and Herring Gull, the male invites the female to "dinner." She frequently responds by frenetically beating her wings, a little like nestlings begging for food from their parents.

In some cases, the male will present his desired mate with prey, such as an insect or a fish. In one experiment conducted by researchers in Europe, a male Corn Crake tried 20 times to copulate with a decoy female placed on its territory. As the pretend female remained impassive before its attempts, the bird improved on its offer by bringing a fat caterpillar.

In place of food, some birds offer a twig. The European Starling even brings a flower from time to time.

The Herring Gull is more "refined." By swallowing, partially digesting and regurgitating his offering to the female, he makes sure that his gift of food will not cause any digestive problems for his chosen mate.

SWINGING COUPLES

Polygamy occurs in 3 percent of bird species and some are even given to "mate swapping." Among them are ratites such as ostriches, emus and rheas; the Hedge Accentor, a small European bird; and tinamous, which range from Mexico to Chile.

In these species, one female will breed with several males or vice versa. Each nest contains the eggs of several females. In the rheas of South America, the male is responsible for incubation.

In Africa, four or five female ostriches may lay up to 50 eggs in a single spot. The dominant female removes some and rolls them into a nearby hollow, where they are abandoned. She may then brood up to 30 remaining eggs, but takes care to place her own eight or 10 at the center of the nest so that they are least vulnerable to predators. In this species, the male takes over the brooding chores at night.

The Atlantic Puffin digs its own burrow in the ground. A colonial bird, it locates its nest by smell. (Page 108)

House Sparrows may copulate for two months before the females lay their eggs. (Page 116)

The egg incubation period in the Blue Tit ranges from 12 to 14 days. This small bird may live for up to 10 years. (Page 121)

The fine layer of down carpeting a Common Eider's nest is gathered while making sure to leave enough for the nestlings. (Page 121)

The Black-legged Kittiwake nests on sea cliffs. Young birds must sometimes leap from great heights to reach the sea. (Page 125)

Although its own diet consists mainly of berries, the Bohemian Waxwing feeds its young insects to ensure that their diet is rich enough in protein. (Page 128)

Life is not always easy in the nests of Snowy Owl. When food is scarce, the eldest nestling may eat the younger ones. (Page 133)

Like the Common Cuckoo in Europe, the Brown-headed Cowbird is a brood parasite. It forces at least 200 other bird species to raise its young by depositing its eggs in their nest. (Page 135)

In monogamous birds, fidelity is more lasting than you might suppose. Even if the male and female spend the winter in different places, sometimes hundreds of miles apart, many pairs find each other again on the breeding territory in spring and raise a new brood together.

But however faithful they may be, pairs generally stay together for only a short period, performing their parental roles until the young have dispersed at the end of the breeding season. However, Brown Creepers and Black-capped Chickadees live with their mates all year long and probably breed with the same partner throughout their entire lives.

In contrast, male and female grouse meet only during copulation. In a number of duck species, bonding often begins on the wintering grounds, but the male leaves the female after copulation.

THE DIVORCE RATE

André Desrochers, a researcher at the Université Laval in Quebec, Canada, studied the Blackbird for several years at Cambridge University in England. He says that for many species, a male and a female will mate with each other throughout their lives. His research on hundreds of subjects showed that 70 percent of males will breed with the same female in the second year. However, given the very high mortality rate, the pair bond rarely lasts long. Moreover, the "divorce rate" rather closely follows the difficulties experienced during reproduction. The larger the brood, the longer lasting the pair, it appears.

These conclusions hold true for other species, as well. For

example, 50 percent of Indigo Buntings keep the same mate each year. The figure is 90 percent in corvids, 68 percent in Merlins and varies from 38 to 54 percent in European Sparrowhawks.

It is difficult to determine scientifically whether pairs manage to spend their entire lives together. However, it is known that species of geese, cranes, penguins, ravens, crows and flamingos have quite durable relationships. One pair of Canada Geese, for instance, lived together for 42 years.

Splitting up is common among Merlins.

THE SECRET LIVES OF BIRDS

The Cost of Housing

After mating, birds turn their attention to housing for their future family. Many birds demonstrate a remarkable aptitude for architecture, engineering and construction; others make do with rudimentary nests. Turkey Vultures are content to lay their eggs on the ground in a crevice or a well-protected spot, which can sometimes be located by the stench coming from it.

Nighthawks also lay their eggs on the ground. Other species lay theirs on bare rock or in a shallow depression. The nests of many shorebirds are merely a small mound of pebbles. Grebes lay their eggs on a mass of floating vegetation. In pigeons and doves, the nest, built of small branches, is often flimsy and fragile.

A large number of passerines build a standard cup-shaped nest. We have even observed American Goldfinches weave their nests so tightly that they are waterproof. Occasionally, Goldfinch nestlings will drown in a drenching rain while their parents are away.

Many types of nesting materials are used. Barn Swallows

make the basic structure with mud. It took one bird 1,200 trips to complete its nest!

House Sparrow nests often contain scraps of cloth, string, plastic and paper. Other species use animal hair or fur.

Nests are often carpeted with fine down and feathers. The Goldcrest will sometimes use 2,000 feathers to create its nest. I have seen a photo of a nest built by a bold American Robin that was made entirely of four-inch (9-cm)-long iron nails. Fortunately, the nest, which contained four eggs, was carpeted with feathers.

DUCKS IN THE TREES

The holes dug by woodpeckers in trees are put to good use by other birds in subsequent nesting seasons. The Pileated Woodpecker, a North American species measuring about 12 inches (30 cm), digs huge holes measuring 6, 8 or even 12 inches (15, 20 or 30 cm). They are so big that ducks can use them once the woodpeckers are gone. Surprisingly, despite their aquatic lifestyle, a number of ducks nest in trees. The most well-known in North America is the Wood Duck. Others include Common and Barrow's Goldeneye, Bufflehead and Common Merganser.

Atlantic Puffins, Leach's Storm-Petrels and European Bee-Eaters dig nests in the earth.

The European Bee-Eater nests in a burrow.

Belted Kingfishers spend three weeks digging their seven-foot (2-m)-long nest hole.

Certain nests are true works of art. Baltimore Orioles, Eurasian Golden-Orioles and a number of European tits weave hanging nests shaped like an elongated pouch. Caciques, colonial species native to Central and South America, build similar but larger nests—up to a yard (1 m) long! When the adult enters, its weight on the bottom closes the hole at the top.

The nest of the Rufous Hornero of South America is quite impressive. It consists of a small mound of earth often built on a fence post or utility pole. Some are open at the front, a bit like little houses, giving the impression of an artificial nest box.

The White Stork and Bald Eagle build nests up to three or even six feet (1 to 2 m) thick. House Sparrows and European Starlings may set up their nests in the resulting tangle of branches. There is one report of a two-ton Bald Eagle nest that finally collapsed during a storm after 30 years of use.

However, the grand prize for architectural achievement surely goes to the Sociable Weaver, a South African species resembling and related to the House Sparrow. These birds build immense collective nests that look like haystacks sitting in trees. A single nest may contain 100 separate compartments, each housing its own nesting pair. Despite the size, ventilation is excellent throughout the nest.

DECOY NESTS

Though most birds build one nest a year, or two or three if they have additional broods, wrens behave quite differently. These tireless builders may be working on up to seven nests simultaneously. In many cases, only one will be used to raise the brood.

The building of such sham nests has long intrigued ornithologists. One theory is that this is a demonstration of sexual prowess by males.

Ornithologist Henri Ouellet, an associate professor at the Université de Montréal, believes that the presence of several nests may also serve to confuse predators. In this way, wrens, most species of which live in the American tropics, may fend off snakes. If the reptile finds nothing to eat in several nests, it may give up—in the animal world, food-finding efforts must produce quick and satisfying results.

Most nests are built within a few days. Exceptions include the nests of large birds like the Golden Eagle. Construction of its nest may take up to two months. But such large birds often merely shore up and reuse the previous year's structure.

The White Stork's nest is voluminous.

Robins take from six to 20 days to build their nests. A Northern Mockingbird can complete its nest in just one day. Woodpeckers may dig for

10 days in the trunk of a tree before finding the hole sufficiently spacious. Other, more particular species take up to one month. The Baltimore Oriole weaves its nest in five to 15 days. Caciques need three to four weeks to concoct their suspended pouches.

The American Goldfinch usually nests in July, and takes 12 days to build its nest. However, when nesting is delayed to mid-August, it will complete the nest in half the time.

NESTING IN THE DEAD OF WINTER

Late nesters such as the American Goldfinch aside, nest building normally takes place in spring. In the south, many species raise more than one brood each summer. The Mourning Dove, the most hunted bird in North America, may nest up to five times a season in the southern United States, starting the last nest near the end of summer.

In the north, climatic constraints force pairs to act quickly as soon as the weather allows, but some "early birds," such as Great Horned Owls, will nest in the dead of winter. The owl often moves into an abandoned crow's nest and begins to lay its eggs in February, when the thermometer reads −13°F (−25°C). Incubation takes 35 days. The chicks' downy coat allows them to survive the cold. The Gray Jay, a relatively tame bird of the boreal forest, nests in March, when temperatures can drop to −5°F (−20°C).

Red- and White-winged Crossbills have, as their name indicates, a very specialized bill. For a good part of the year, they live on seeds, which they extract from conifer cones. When food is scarce in any given area, they move on to the next source. It is thought that their nesting period observes the dictates of food availability, for in North America these birds breed from January to April.

In Europe, where crossbills are more numerous, the nesting period varies according to habitat. In hemlock forests, the birds brood in late summer and early fall. In large pine forests, the activity takes place in spring, whereas in spruce forests, it occurs from fall to early winter. In mixed conifer forests, breeding may take place almost year-round.

The Rock Dove may raise two or three broods at different periods of the year, even in winter. Several years ago one nestling fell off of a disused awning that was moved in late January in Canada, when the thermometer read –5°F (–20°C).

Eggs: A Full-Time Job

In bad weather, many passerines delay egg laying by several days—even up to a week for Tree Swallows. In most cases, the time between fertilization and egg laying is 24 hours, but it may extend over several days.

An egg begins its existence as an *ovum* produced by the left ovary, the only functional one in birds. Numerous and of varying sizes, depending on their stage of maturity, these yellow ova are released one by one into the bird's *oviduct*.

The actual laying process usually takes only a few minutes, although in turkeys it can last over two hours. Parasitic species like the cuckoos and cowbirds lay their eggs very quickly in the host bird's nest—sometimes in just a few seconds—taking advantage of the prospective foster parents' brief absence.

Becoming an Egg

After it matures, the ovum is released into the *infundibulum*, a chamber where it remains for about 20 minutes; it then enters the magnum, where it will be coated with albumen. From there, it slowly travels down the oviduct, where the albumen is wrapped in its shell membranes. Then it is on to the uterus, where the shell, composed almost entirely of calcium carbonate, is added. At the conclusion of the process, the egg is expelled by muscular contractions and a new ovum begins to mature.

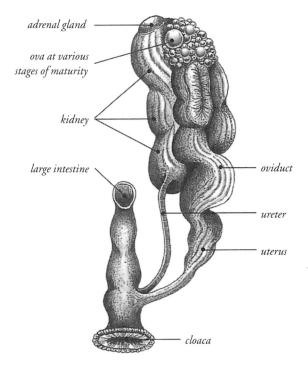

adrenal gland

ova at various stages of maturity

kidney

large intestine

oviduct

ureter

uterus

cloaca

REPRODUCTIVE SYSTEM

As in mammals, calcium is an essential element of a bird's diet, particularly during the breeding season. Birds get their calcium from water, the stones used as abrasives in the gizzard, animal and vegetable food sources, mollusk shells and eggs.

Even if a bird's diet were completely devoid of calcium during the laying season, the reserves in its bones would probably be enough to produce a clutch of eggs. Calcium is stored in the bone marrow and about one-fourth of the available supply is used in egg laying. In domestic chickens, this quantity is sufficient to produce a half-dozen eggs.

When an egg is laid without a rigid shell, the problem is not always a calcium deficiency. It may be a hormonal disturbance, or the contractions of the oviduct may be too quick and violent to give the calcium time to form around the ovum.

It is now known that the presence in a bird's diet of certain pesticides, such as DDT, may modify the egg's chemical composition and affect its solidity. Contaminated eggs may break under the weight of the parent bird.

In the 1960s, North American Brown Pelican and Peregrine Falcon populations underwent a catastrophic decline. Research revealed that, over long periods of time, these birds had eaten prey whose flesh contained DDT residues. The U.S. and Canadian governments reacted by prohibiting the use of this toxic product, and since then the affected populations have begun to recover.

THIS WON'T TAKE BUT A MINUTE

When ready to breed, the male presses its *cloaca* (an orifice that is both genital and intestinal) against the female's. It maintains this precarious position by hanging on to a clump of feathers on its mate's neck or head.

Depending on the species, copulation can take place in the water, on the ground, in the nest, on a perch, or even in flight, as is typical of swifts.

Although sometimes attended by lengthy rituals, copulation itself typically lasts just a few seconds. Naturally, there are exceptions. The Aquatic Warbler, a small European bird, invests considerable time and energy in the sex act. The male stays attached to the female's body for 20 minutes at a time, ejaculating repeatedly.

Most birds do not have a penis, but certain species such as ostriches and other ratites, ducks and game birds have a small one.

Birds copulate many times before and during egg laying. House Sparrows copulate for two months before laying. The Black Duck, a species that nests in the boreal forests of eastern Canada, begins copulating as early as December. However, there is no exchange of sperm at that time because the sexual organs do not reach maturity until April.

After ejaculation, the sperm travel up the oviduct to the ova, a process that may take only 30 minutes, although fertilization takes somewhat longer. The sperm may remain viable for several weeks, so that most birds can produce fertile eggs well after laying the first egg. Females of some species even have a sort of "sperm bank" that can be "drawn on" for up to two months after copulation.

Typically, a fertile egg is laid about three days after copulation, although this period is reduced to 24 hours in domestic chickens. A vast quantity of sperm pass from male to female during copulation—more than three billion in a rooster's semen, for example.

Egg Beaters

Clutch size is invariable in some bird species but not in others. Clutches of from one to 23 eggs have been recorded. Many birds lay more than one clutch a year.

Under artificial conditions, birds can be made to lay many eggs. In one oft-cited turn-of-the-century experiment, researchers removed each egg of a Northern Flicker, a familiar North American woodpecker, as soon as it was laid. The bird kept right on producing—71 eggs in 73 days!

Canadian ornithologist André Desrochers reports that, in England, a female Blackbird whose eggs were repeatedly stolen by a squirrel produced eight clutches, totaling 30 eggs, in one summer. A Mallard duck will replace her 12-egg clutch if the first one is destroyed before hatching.

The number of eggs varies according to the female's health, the quantity of available food and climatic conditions. Selected for their performance, certain domestic hens produce an average of 274 eggs a year.

Producing an egg is very energy-intensive. For example, the egg of a tiny wren represents 13 percent of its body weight. The figure is less than 2 percent in ostriches, but it reaches an incredible 23 percent in kiwis. The Brown Kiwi lays two or three eggs a year at four-week

One Northern Flicker laid 71 eggs in a summer.

intervals, each weighing about a pound (500 g). At the other extreme, the smallest hummingbird egg weighs barely 1/100 of an ounce (0.2 g).

BIRTH CONTROL

Egg production can be affected by climatic conditions and food abundance. For example, on arriving at their Arctic breeding grounds, Snow Geese depend entirely on their stored energy reserves to lay four or five eggs.

In 1992, the Snow Goose population of eastern North America, which passes through the Saint Lawrence basin during migration, was unable to nest due to the harsh climatic conditions prevailing on northern Baffin Island. When the geese arrived in late May, winter had not yet ended, and in mid-June snow and ice still covered much of the breeding grounds. Even if the geese had waited for suitable conditions, their young would not have been sufficiently developed to fly south before the first frost, around August 20.

Kiwi eggs weigh one-fourth as much as the bird itself.

Food availability is also vitally important. The number of eggs laid by the Snowy Owl is often directly related to the size of the lemming population on its territory. When food is scarce, it does not nest at all.

Many songbirds raise more than one brood per year. Two broods are common in the Barn Swallow, European Starling, European Robin, Blackbird and American Robin. In general, the second brood is smaller and its survival rate is lower.

The record holder for the number of broods per year—up to 21—is probably the Zebra Finch, a small passerine living in the semidesert expanses of Australia.

LONG-LIVED LAYER

At what age do birds begin to lay eggs? In some species, breeding begins within a year of their own hatching—just eight weeks later in House Sparrows. By contrast, large raptors and gulls must wait several years before raising a family. The Wandering Albatross does not breed before the age of seven.

In biology the rule is, the smaller the animal, the less time it lives, the more often it breeds and the more offspring it must produce in order to ensure the survival of its species.

On the whole, birds live short lives, but some specimens raised in captivity have reached venerable ages and had remarkable reproductive exploits.

The Wandering Albatross takes seven years to reach breeding age.

The story of two Eurasian Eagle-Owls raised at the end of the last century in England is well known to ornithologists. The male and female were still copulating at the ages of 53 and 68, respectively. The female laid eggs for 32 years and brooded 93 nestlings.

One captive Herring Gull bred for 42 years. The record is 32 in the Canada Goose. There is also the unusual case of a Rock Dove that was still laying eggs at the venerable age of 30, although the species rarely survives more than eight years in the wild. In New Zealand, a Royal Albatross produced eggs for 48 years, not in captivity but in its natural environment.

AN EGG A DAY

Birds can lay no more than one egg per 24-hour period. This frequency is the rule in a wide range of species, including songbirds, ducks, shorebirds and woodpeckers.

At the start of the breeding season, birds normally lay an egg every two days, a little longer for some raptors, large waders and hummingbirds. However, this interval is three days for the emus of Australia and for puffins, five days for the Greater Spotted Eagle and eight days for Mallee fowl. Mallee fowl lay their eggs in compost that serves as an incubator.

Most birds begin to brood their eggs only after laying is complete so that all of them hatch simultaneously. But some birds, such as owls, and especially the Snowy Owl, begin to brood as soon as the first egg appears. This can produce conflicts among nestlings when there is a large age difference between them.

Incubation time varies from one species to another. It tends to be brief in small species and longer in larger ones. For instance, the shortest incubation period is 12 days, but the Royal Albatross broods for 12 weeks.

Sample incubation times are 12 to 14 days in many birds, including House Sparrows, Blue Tits, several woodpeckers and American Robins; 13 to 15 days in Blackbirds; 17 to 19 days in Rock Doves; 25 to 27 days in Herring Gulls; 26 to 30 days in Mallards; 35 days in Bald Eagles and 43 to 45 days in Golden Eagles.

THE BROOD PATCH

In general, females stay on the nest and males keep watch and sometimes bring food. But in the phalaropes and some other birds, the roles are reversed.

Although parent birds are motionless while brooding, the activity requires a large amount of energy. But nature does its work well. In many species, particularly those with *altricial* young (where the chicks are born completely dependent on the adults and remain in the nest until they can fly), the brooding bird develops a patch of bare skin called a *brood patch*, which provides better contact between the skin and the eggs. The heat transferred from the parents' bodies to the eggs through a multitude of tiny blood vessels demands a significant outlay of energy. It is estimated that brooding accounts for 16 to 25 percent of a bird's daily energy expenditure. The patch slowly disappears after hatching.

Species lacking a brood patch use efficient methods to warm their future progeny. Ducks and geese carpet their nest with a highly insulating down (in fact, eiderdown is gathered from these nests). Some birds lay their eggs on their own feet. Penguins insert them into a fold of their skin.

THE EGG TOOTH

As brooding progresses, the embryo develops, causing a complete alteration in the egg's structure. The many pores in the shell (more than 7,000 in a hen's egg) allow water vapor, respiratory gases, even micro-organisms to pass through. Gradually, the egg loses water and the shell becomes thinner as the calcium is drawn off to make the embryo's skeleton. During incubation, the egg loses up to 20 percent of its weight.

Shortly before hatching, the chick pierces the membrane, creating an air pocket at the wide end of the egg, and begins to breathe air. The chick begins calling at once and can easily be heard. It is sensitive to goings-on outside the shell and reacts to the adult's calls.

With its powerful neck muscles, the chick alternately presses against the shell and rests. This requires considerable energy. The chick uses a protrusion on the tip of its upper mandible known as an *egg tooth* to crack the shell and get free. The egg tooth falls off after serving its purpose.

In *precocial* birds (ducks, geese and chickens, for example), the entire brood hatches almost simultaneously. Shortly afterward, all the chicks are able to leave the nest because they are fully formed at birth. They are covered with warm down and able to feed themselves.

Hatching usually takes a few hours, but with the Winter Wren it lasts two days. In the pelagic Sooty Shearwater, chicks take four days to emerge from their shells. When altricial chicks hatch, parents eat the empty shells or dispose of them so as not to attract predators. It has been shown that nesting success decreases where shells remain in the nest.

Chicks of some species call for food almost immediately, but others may be silent for several hours. This respite benefits the parent birds, since another exhausting phase of the process is about to start: feeding the young.

Pintail chicks are precocial.

Feeding the Family

The first steps of precocial chicks are fraught with peril. Before facing the outside world, tree-nesting ducks perform an acrobatic jump from nests as high as 55 feet (17 m). This is the case for Wood Ducks as well as goldeneyes and mergansers. Nevertheless, the chicks usually land comfortably on the ground and injuries are rare.

Ducklings often must walk hundreds of yards, even half a mile or a mile, to reach water, where they can eat their first meal and rest. Along the way, however, predators are on the lookout for them.

A number of seabirds nest on high promontories such as sea cliffs. Their chicks may have to leap from a height of more than 325 feet (100 m) to reach the sea. If their runway is poorly located, the birds may crash.

Familiar ducks like Mallards and American Pintails often nest on agricultural land near the suburbs. There, too, the young birds encounter obstacles. For example, they must cross roads and highways to reach the tranquillity of the ponds. Fatal accidents are numerous.

Another peril facing precocial birds, particularly grouse, is climate. Cold or persistent rain in the days after hatching can wipe out an entire brood. Precocial birds are not fed by their parents, but the parents do locate feeding areas for them and look after their security. Female Ruffed Grouse are commonly seen attempting to draw away intruders by faking an injury. The distraction gives the chicks time to take cover.

The parents also monitor the quality of the food their young eat. Domestic hens select seeds for very young chicks and deposit them one by one on the ground before them. This behavior continues until the brood has become autonomous.

AN EXHAUSTING TASK

In altricial birds such as robins, chickadees and wrens, the parents' work is much more exhausting.

The chicks are more poorly developed than those of precocial birds and unable to satisfy their own needs. Many are born scrawny and naked and keep their eyes closed for several days. However, they are able to crane their necks and receive food shortly after hatching. By the time they leave the nest, they will have grown to almost the size of their worn-out parents.

If the adults did not rest at night, they would literally kill themselves looking after their young. For example, the House Wren, a bird weighing little more than a third of an ounce (10 g), works tirelessly raising its brood of seven. One pair made 665 trips—most of them by the male—during a 65-hour period to feed the young. Researchers also watched one male, which had no doubt lost its partner, make 1,217 trips during the daylight hours, or one every 47 seconds, to feed its 12 day-old young. As if that weren't enough, House Wrens raise two, sometimes even three, broods each summer.

In the Antarctic, Chinstrap Penguins make 200 dives a day to find shrimp for their young. The Belted Kingfisher captures about a hundred four-inch (10-cm) fish each day to feed its six or seven chicks—a considerable task, given that the success rate on each dive is only about 45 percent. Common Swifts fly 550 to 625 miles (900 to 1000 km) a day to meet the needs of their two or three nestlings, sometimes bringing back hundreds of insects on a single trip.

However, it is estimated that feeding frequency is four to 12 trips an hour on average, depending on the species. European Starlings may feed their young up to 17 times an hour. Swallows make 30 trips during the same period. Mealtimes vary but are usually more frequent in the morning, when the chicks are at their hungriest.

FAST GROWTH

Chicks are extremely demanding. Those of the European Starling

Common Swifts fly 550 miles (900 km) per day in search of food for their young.

eat nearly the equivalent of their own weight every day. Belted Kingfisher young are even more demanding. Growing American Robins eat more moderately—only half their weight every day. Still, the daily intake of worms by one young robin would measure more than 20 feet (6 m) long when placed end to end.

Some seabirds, hummingbirds, large waders and crossbills predigest food for their young by swallowing it, carrying it to them and regurgitating it into their bills.

The diet is particularly rich during nesting to support the fast development of the young. Certain parents even enrich the daily diet with fragments of bone or shells to provide the calcium needed to form bone. Adult House Sparrows, themselves almost exclusively vegetarian during this period, nevertheless feed their young mainly insects.

Compared with those of mammals, growth rates of young birds are often phenomenally fast. At three weeks, a Common Cuckoo is already 50 times larger than at birth. Seven weeks after hatching, the weight of a Great Blue Heron will have increased by a factor of 38.

The Snow Goose is also interesting in this regard. When the chick hatches in the Arctic in early July, it weighs about three and a half ounces (100 g). Although the Arctic summer is short, the daily 24 hours of daylight give the birds a chance to eat all day long, stopping only briefly to digest.

Six weeks later, the young geese weigh at least two pounds (1 kg) and often twice as much if conditions are favorable—or 20 times their weight at birth. Assimilation and transformation of food occur at a rate that is probably near the biological maximum for birds.

Young Common Eiders (here, an adult male) are often raised in "nurseries" containing many ducklings that are watched over by a dominant female. (Page 141)

The Semipalmated Plover, a shorebird, migrates by night. Some will die in collisions with lighthouses. (Page 147)

The first migration of many Snow Geese juveniles will be their last. Nearly 90 percent of geese killed during the hunting season are first-year birds. (Page 132)

The Arctic Tern holds the record for the longest annual migration. (Page 149)

One Laughing Gull reached the venerable age of 63 in the wild. (Page 154)

A banded Mallard lived to 29 in nature. One pair (female at left) may produce a dozen ducklings each season. (Page 154)

Usually wary, the Ruffed Grouse may sometimes, as here, lose its fear of humans. (Page 158)

Bird behavior can vary from one location to another. Although at ease in English gardens, the European Robin tends to be cautious and secretive in the forests of France. (Page 156)

Considered to be the most abundant bird in North America in the nineteenth century, the Passenger Pigeon was completely wiped out by intensive hunting and destruction of its habitat. (Page 165)

Good climatic conditions also play an essential role in nesting and chick survival. In altricial species, parents spend considerable amounts of energy carrying out their mission. Their frequent movements make them more susceptible to predation, and mean they must eat more to accomplish their tasks. If the air temperature drops, parent birds must spend even more energy to ensure their own survival. Among insect eaters, the situation is complicated by the fact that insect hatching and activity decrease as the temperature drops. The young birds suffer the consequences.

In Eastern Canada, Tree Swallows arrive at their breeding grounds in late April and are at their nests as of early May, when the temperature is still subject to sharp fluctuations. Rain and cold (the temperature is sometimes just above freezing) may destroy several broods because of lack of food. Adults may die for the same reason.

HOUSEHOLD CHORES

Parents also have other tasks. They must keep their chicks warm. In Europe, young Common Swifts can survive their parents' absence by falling into a state of torpor. If it rains or the sun is too strong, the parent bird may spread its wings like an umbrella to shield the young.

Birds usually get the water they need from their food. The case of the sandgrouse, a species reminiscent of the Gray Partridge, is unique. This bird lives mainly on arid steppes, where water is scarce. The male's breast feathers are structured to absorb water—much like a sponge.

After removing the oils from its feathers, the bird submerges itself in water for several minutes, then returns to the nest up to several dozen miles away. The Namaqua Sandgrouse, for example, a resident of the Kalahari Desert in Africa, may travel up to 50 miles (80 km) in search of water. Although some of the water evaporates along the way, the young manage to draw enough from the adult's feathers to quench their thirst for a good part of the day.

Parent birds also keep the nest clean and protect it from predators. In most passerines, the excrement ejected from the nest is wrapped in a membrane called a *fecal sac*. In some cases, the parents swallow this sac, which may contain partially undigested food from the chicks.

Most passerines clean their nests constantly and will tolerate no foreign bodies. On one occasion, the parents even threw out a nestling in an effort to get rid of a band attached by a biologist to its leg. The parents realized their mistake and brought the bird back into the nest.

Young raptors have an innate reflex to answer nature's call without soiling the nest. They perch on the rim so that the excrement falls outside. With herons and other large waders, the nest structure is so loose that the fecal matter passes straight through.

The House Finch, on the other hand, and certain raptors such as the Great Horned Owl seem unconcerned by matters of hygiene, and their nests may be truly filthy.

Special feathers allow the male Sandgrouse to carry water to its young.

AND THEY'RE OFF!

Typically, the young spend about as much time in the nest as inside the egg. When the time is right, the young birds depart. In most cases, all the chicks leave the nest at once. Sometimes, when there are stragglers, the parents won't bring as much food, to encourage them to fly or force them to leave the nest to obtain their own food. In other cases, the chicks are simply pushed out.

In species that raise more than one brood, the female builds the second nest by herself and begins to lay the eggs while the male takes charge of the first brood.

In many cases, the parents continue to feed their young even after they have left the nest. Most altricial species do so for several days or weeks, until the young are completely autonomous. In terns, feeding continues for several months as the parents teach the young to fish. Several raptors behave similarly. On the other hand, young Common Swifts and many seabirds are independent as soon as they leave the nest.

Although the Snow Goose, crows, the Black-capped Chickadee and some others flock with their parents until the next breeding season, the young of most species must learn to survive on their own.

The offspring of nonmigratory species must seek a new territory. Adult males do not hesitate to chase off the young, which they now perceive as competitors.

Adult life brings with it a new set of problems. Mortality is highest for first-year birds, and many newly fledged individuals, unable to feed themselves adequately, will die within the first few weeks.

The Snow Goose, a highly prized game bird in eastern North America, is particularly vulnerable during the fall hunting season. It is estimated that 90 percent of geese killed at that time of year are first-year juveniles, who are inexperienced and unaware of the threat posed by hunters. But once birds have survived their first hunting season, they have a good chance of growing old.

Predators and the hazards encountered during fall and spring migrations are also major causes of mortality. It has been found that often only 10 percent of young birds will return to breed the following year.

FRATRICIDE AND CANNIBALISM

Life in the nest is often eventful—and sometimes provides drama of a high order. The law of survival of the fittest is as true for nestlings as it is for the rest of nature. Family conflicts are numerous and occasionally end tragically.

Cases of fratricide and cannibalism are frequent. In some raptors, sibling rivalry begins as soon as the eggs hatch. The first chick out quickly establishes dominance over the others and its priority access to food. It will readily kill its brothers and sisters to maintain its supremacy.

In Golden Eagles, the older, generally more developed, nestling often attacks and kills its younger nest mate. In Bald Eagle broods, which usually consist of two or three chicks, the aggressions of the eldest sibling tend to tail off after the first few weeks. At that point, the younger birds—assuming they have survived—can usually begin to lead a peaceable existence.

Brown Booby clutches normally consist of two eggs. The

younger, weaker bird is often ejected from the nest, and the older bird seizes the opportunity to make sure it doesn't return. Lacking food, the young bird will die after some time. The same family dynamic occurs with American White Pelicans and Whooping Cranes.

The situation is no less turbulent for Snowy Owls. They lay four to eight eggs at irregular intervals, and up to 14 when food is abundant. Incubation lasts about 33 days. The first nestling will sometimes leave the nest even before the last one has hatched. When food is scarce, however, the famished older nestling will sometimes eat the youngest of the brood. This phenomenon has been observed in Great Horned Owls and other raptors, as well.

CANNIBALISM— IT'S ALL IN THE FAMILY

Many bird parents will eat their dead or sickly chicks. This type of cannibalism occurs in species such as the White Stork, Red-backed Shrike, Barn Owl and diurnal raptors like the American Kestrel.

In times when food is scarce, Barn Owl parents will kill and eat some of their offspring, beginning with the smallest and weakest. So do the Greater Roadrunners of cartoon fame. It is believed that, by doing this, parents enhance the remaining chicks' chances of survival.

Cannibalism is particularly prevalent in colony-nesting birds. In gull colonies, by the time the young are ready to fledge, there are often fewer chicks than there are breeding pairs. These losses are due partly to predation by other birds, but mainly to cannibalism, which occurs frequently when the colony is disturbed.

Indeed, the slightest disturbance in the colony may result in significant chick losses, because it drives chicks out of their families' territories. In high-density colonies, each pair's territory is usually limited to just a small space around the nest. Despite the general chaos that appears to reign in the colony, strict social and territorial systems are at work. Chicks that leave their parents' territory, whether out of fear or simply out of carelessness, run the risk of being attacked and eaten by neighbors. In Herring Gull colonies, chicks that stray from home but manage to return safe and sound may no longer be recognized by their parents, and will quickly be killed and eaten.

The Greater Roadrunner sometimes eats its offspring when food is scarce.

THE SECRET LIVES OF BIRDS

Finding a Full-Time Babysitter

Brood parasitism is another problem that plagues some species. Following instincts that have evolved over many millennia, parents will put food into the chick's mouth that is open the widest. Since parasitic nestlings are generally larger than the host species' chicks, in extreme cases parents will continue to ply the adopted nestling with food when their own chicks are starving to death.

Obligate parasitism, in which the parent birds rely completely on a host species to raise their young, occurs in cowbirds, honeyguides and some cuckoos. The Brown-headed Cowbird in North America and the Great Spotted and Common Cuckoos in Europe are incapable of passing on their genes without parasitizing another species' nest.

Other species, called *nonobligate parasites*, occasionally engage in parasitism. Mergansers and pochard ducks, as well as House Sparrows, European Starlings and some woodpeckers, sometimes lay their eggs in other birds' nests. Ornithologists estimate that close to 900 species in the world fall into this category.

Nonobligate parasitism is fairly common in colony-nesting birds, as well as in ducks. Among the 20 or so species of ducks that engage in parasitism, Redheads and Ruddy Ducks are particularly frequent parasites; some ornithologists believe that they are evolving toward obligate parasitism. One South American species, the Black-headed Duck, has already reached this point and always lays its eggs in other birds' nests.

In North America, probably our best-known woodpecker, the Northern Flicker, sometimes parasitizes the nests of Eastern Bluebirds, Tree Swallows and Pileated Woodpeckers.

Ring-necked Pheasants occasionally lay their eggs in the nests of other species of pheasants, domestic chickens, Ruffed Grouse and Northern Bobwhites. Bobwhites also use Ring-necked Pheasants as hosts. In Canada, House Finches, Brown Thrashers, Virginia Rails and Eared Grebes take a break from parenting in this way from time to time.

The foster parents raise the adopted young just as they would their own offspring, as in the case of a pair of House Sparrows that reared a Bank Swallow nestling, brooding and feeding it along with their own chicks. Sometimes an unsuitable host species is chosen, however. The Common Shelduck, for example, an Eurasian diving duck that usually forages in shallow water for plants and invertebrates, sometimes lays its eggs in the nest of a Red-breasted Merganser, which dives deep for its prey, almost always fish. The adopted Sheldrake ducklings do not survive for long in this situation.

THE CUCKOO AND THE COWBIRD

Cuckoo and cowbird chicks often manage to do away entirely with their competition, the foster parents' original brood.

The young of the Common Cuckoo (and its cousin, the Great Spotted Cuckoo) hatch nearly naked and with their eyes closed, yet they somehow push all the other eggs out of the nest with their tiny wings. Honeyguide nestlings kill the host family's chicks outright, using the sharp hooks on the tips of their bills. Cowbirds have a less aggressive but equally effective strategy: to gradually usurp all the space available in the nest. Despite the irksome presence of the intruder, host parents sometimes succeed in raising some of their own young to fledging.

The type of parasitism found in birds is almost unique in the animal kingdom, except for some insects. Scientists believe that it is a form of predation but are not sure what has caused species to evolve in this direction.

Common Cuckoos lay their eggs in other birds' nests.

Parasitic females in search of foster parents closely monitor the parents' comings and goings from the nest, and as soon as it is unguarded they hurry in and lay their eggs. Cuckoos usually remove one of the host's eggs and replace it with their own, often in just a few seconds. Brown-headed Cowbirds, on the other hand, simply lay one or two eggs amid the original clutch, usually producing 10 to 12 eggs a season but sometimes twice or three times as many. Cuckoos generally lay a dozen eggs, but occasionally may produce up to two dozen.

THE FIRST TO HATCH

As a rule, the egg of the parasitic bird is the first to hatch, and the young cowbird or cuckoo immediately begins to cry for food. When grown, the parasites are usually bigger—up to three times bigger—than their hosts. The other nestlings have to make do with leftovers; they are unlikely to be properly fed, especially if food is scarce. At times, young birds will be tossed out of the nest to make more room for the parasite.

In North America, the Brown-headed Cowbird uses at least 200 bird species as foster parents, but its favorites are the Yellow Warbler and the Song Sparrow. The Common Cuckoo tends to seek out the nests of the Great Reed Warbler and the Bull-headed Shrike.

The adaptive characteristics of parasitic birds are remarkable. The Common Cuckoo begins to incubate its eggs even before laying them—a tremendous advantage for the survival of the species.

Another phenomenon is *egg mimesis*. It is so refined in cuckoos that even experts have great difficulty telling the host's egg from the parasite's. In one of 60 African parasitic cuckoo species, the resemblance was so close that it took a genetic, or DNA, analysis to figure out which egg belonged to whom. In Finland, the Common Cuckoo's eggs are bluish like those of its typical hosts, the Common Redstart and Whinchat. If the cuckoo parasitizes the Great Reed Warbler, the eggs will be greenish with black spots—again just like the host's.

Cuckoo eggs are normally quite robust, another evolutionary advantage for this parasite. The bird sometimes lays its eggs by dropping them from a considerable height, and their hard shells may even break the eggs already in the nest.

The impact of parasitism varies considerably, but host populations on the whole do not seem to be gravely affected by it. In certain areas, though, parasitic birds have caused a significant decline in their host species. The task of raising one additional offspring is very demanding. If the parents are exhausted by the effort, they are unlikely to raise another brood.

Some species succeed in fending off the intrusions of parasitic birds. In North America, the American Robin and the Gray Catbird seem able to recognize cowbird eggs and eject them from the nest, although the combativeness of robins varies from one location to another.

Some species desert their nests as soon as they discover an unwanted egg in it. Others, like the Yellow Warbler, build a new floor on top of the parasite's egg and start over. Research has shown that they will do this up to five times on a single nest if the parasite is particularly persistent. This instinct in the host is usually strongest before it finishes laying its eggs, when there are only a few in the nest. If the cowbird arrives when laying is almost complete, the warbler tends to accept the new egg, incubating it along with its own.

When the host parents have completed their child-rearing tasks and the young cowbird has become fully independent, it instinctively seeks out members of its own species.

HELPING THE NEIGHBORS

Unlike parasites, birds of many species will help individuals of their own species, sometimes even a neighbor of a different species. Recent research has shown that this phenomenon is much more widespread than previously believed, and occurs in at least 200 species.

The Florida Scrub-Jay provides a good illustration of mutual aid. On occasion, four or five jays that are too young to breed provide support to parent birds. They help them feed the young, clean the nest and protect it against intruders. In some species, a helping attitude runs in families. For example, first-brood Eastern Bluebird nestlings sometimes help their parents feed the second brood.

Australian scrubfowl have a very peculiar style of communal incubation. Several scrubfowl females will lay their eggs in an insulating layer of plant debris; the heat given off by the decomposing organic matter causes the eggs to hatch. These scrubfowl nests can be up to 33 feet (10 m) in diameter and 16 feet (5 m) high. The males of this turkey-sized bird constantly monitor the nest temperature with their bills, adding or removing matter as necessary to adjust the temperature. Egg laying can take place over much of the year. Males spend four to five hours a day monitoring the site. The eggs take 50 to 90 days to hatch. Nestlings must then make their way through the pile of debris in order to see the light of day. They can run from the minute they hatch, and can fly 24 hours later. Independent from birth, they live solitary lives except during breeding periods.

In the southern United States, Mexico and Central America, mating pairs of the two species of ani, black birds with peculiar bills, work together building nests, incubating the eggs and feeding the young.

TAG-TEAM BROODING

Even more unusual breeding behavior is known to occur. In one case, an American Robin and a Cedar Waxwing brooded their eggs together in the same nest. The waxwing sat on the eggs, while the robin sat on the waxwing, or vice versa. Communal life lasted until the eggs hatched, at which point the parents each looked after their respective broods.

In another instance, a nest was shared by a Red-shouldered Hawk and a Barred Owl. The hawk brooded at night while the owl was out hunting. John K. Terres also reports the behavior of a male Carolina Wren carrying food to a brooding female House Wren. After the eggs hatched, the adoptive father was so intent on feeding the young that the real parents let him do all the work.

Henri Ouellet, who headed the ornithology department at the Canadian Museum of Nature for several years, observed a similar scene on the tundra. A Savannah Sparrow insisted on feeding the young of a Lapland Longspur pair, despite their continued efforts to chase it away from the nest.

Ouellet believes that some birds behave this way in order to use up their instinctive energy when their own brood has been destroyed, for example. In other cases, their urge to feed young birds continues even after their own have become independent.

NURSERIES FOR EIDERS

Quite large groups of Common Eider ducks can be observed along the Atlantic coast or in the Saint Lawrence estuary. Too big to be individual families, they are actually a kind of nursery. This commonest of the eiders is a colonial nester. Often, several dozen clutches of three to five eggs hatch at the same time. If the timing is right, the young quickly leave the nest for the sea in the company of the female. On their journey, the family must constantly fend off hungry predators such as Herring Gulls and Great Black-backed Gulls.

Great Black-backed Gulls prey on young Common Eiders.

In the ensuing chaos, ducklings may wind up in other families or with a group of several dozen young birds. A number of females, including some with no young, then take these birds under their wing, so to speak. Although adoption is quick, harmony in the new family is short-lived. The females quickly establish a hierarchy in which the stronger ones chase away the weaker. After a few days, only one or two females will have taken charge of the whole group of fledglings, keeping any other parentally inclined adults at a distance. One adult duck was observed with 112 ducklings in its care.

Penguins, pelicans and some tern species sometimes raise their young in the same way.

THE STRANGE CASE OF THE HORNBILL

Of all birds, the hornbills have the most remarkable nesting technique. Up to 5 feet (1.5 m) long, these African and Asian birds are equipped with a large bill reminiscent of the toucan's.

However, hornbills sometimes have a large protuberance on the upper mandible.

Hornbills mainly nest in the large hollows of trees. Before settling down in the nest, the female starts closing the hole with mud, leaving enough space to enter. Once inside, she literally walls herself in by blocking the entryway with excrement or food. In some species, the male assists her.

From this moment on, only the bill can emerge from the hole. The male thus becomes the sole purveyor of food, a role he will fulfill until the brood fledges some 20 to 40 days after hatching, depending on the species.

In some cases, the female breaks out of the nest to help her partner feed the one to seven nestlings, but as soon as their mother has left, the young birds hurriedly seal up the hole.

This solitary confinement serves to safeguard the family. The nest usually has an escape hatch for emergencies.

This female Abyssinian Ground-Hornbill is one of the few that does not shut herself up in her nest.

THE SECRET LIVES OF BIRDS

Migration:
A Perilous Journey

Even today we still do not fully understand why birds migrate in spring and fall. At first, the answer seems obvious: they are forced to leave to avoid the cold and snow that will reduce or eliminate their food sources.

However, the migration process varies considerably from one species to another and many behaviors remain unexplained. Why are there migratory and sedentary members of the same species in the same region? Why are there vast differences in migration distance among birds of the same family? Why do some travel thousands of miles, when much closer, ostensibly suitable wintering sites are available?

Migratory patterns are not immutable. Over the past several years, many Canada Geese in eastern North America have stopped migrating to northern Canada in spring. Instead, they live year-round on their wintering grounds in the United States. In many places, the geese have become real pests in public parks and on golf courses because their droppings cause serious public health problems.

A large number of Barn Swallows leave Canada to winter in South America, and some have begun to nest in Argentina instead of returning. In Europe, the Serin is gradually extending its range northward out of the Mediterranean region. As well, formerly sedentary southern populations have become migratory by moving north. In the past few decades, many Blackcaps in Germany have adopted the British Isles as their winter refuge instead of heading for the Mediterranean.

The warbler-sized Northern Wheatear is found worldwide. European populations nest on the tundra and spend the winter in West Africa. The species began nesting in North America a few decades ago. Birds now breed in northern Canada, which involves starting out from Africa, then crossing Europe and the North Atlantic by way of Iceland and Greenland.

At the end of the summer, they make the trip in reverse. Researchers believe that Northern Wheatears nesting in northern Canada could eventually begin migrating to South America for the winter, just as the Barn Swallow did at some time in the past after arriving in America from Europe.

TREACHEROUS TRAVELS

According to Frank B. Gill, about 10 billion individual birds go on migration each year. Half of them (187 species) leave Europe and Asia for Africa; the others (more than 200 species) travel from Canada and the United States to Mexico, Central and South America and, in some cases, the Caribbean.

Migration demands incredible endurance and power. Birds encounter many perils and some die en route. As Gill observes, of the 100 million ducks and geese that migrate south each winter, fewer than half return to breed in the spring. As well, it is estimated that half of the first-year migrants from all over the Northern Hemisphere never return from their journey.

Predators, storms, exhaustion and accidents cause a considerable number of deaths. Entire bird flocks have perished at sea due to disastrous climatic conditions. Spectacular accidents may also occur. From time to time, flocks of birds collide head-on, and each year thousands of birds meet their deaths by crashing into buildings or tall structures.

Despite these hazards, migration has distinct advantages for the species that undertake it. Researchers have found that the mortality rate for year-round temperate-zone residents is higher than that of species migrating to the tropics. However, the sedentary populations have larger broods.

How did birds first begin to migrate? Ornithologists agree that birds are opportunistic, and the annual migration is a means of improving their quality of life. According to one theory, birds may have gradually extended their breeding range north during a period of global warming. At winter's approach, being intolerant of the cold, they may then have instinctively returned to their original breeding grounds.

Another more evolution-centered theory holds that birds were already nesting in the North at a time when the climate was warmer, and that vast glaciations may have forced many of them to leave the region in late summer. Only the hardiest of those that did not migrate would have survived.

NAVIGATION BY SUN AND STARS

The urge to migrate is touched off by *photoperiodism*, a hormonal response in birds to diminishing daylight in the fall. This phenomenon allows the animal to prepare for departure by building up fat under the skin.

Birds use a variety of cues to guide them on migration, including the position of the sun or stars, the Earth's magnetic field, the prevailing winds and the topography. In Europe, the Rock of Gibraltar is a major gathering point for birds crossing the strait to Africa.

In May, thousands of raptors and a host of warblers and other passerines rest at Point Pelee National Park in Ontario, Canada, after crossing Lake Erie. Each year, thousands of amateur ornithologists are on hand to observe them.

The sense of orientation in some birds is nothing short of astonishing. In the 1950s, researchers conducted an experiment by removing a Manx Shearwater from its nest burrow on an island off the coast of Wales, flying it to America and releasing it immediately on arrival. The shearwater, a pelagic, or ocean-going, bird, was found at its point of origin about 12 1/2 days later. To get home, it had had to travel 3,100 miles (5000 km) across the Atlantic Ocean at a rate of 250 miles (400 km) a day.

The Golden Plover nests in the Arctic and spends the winter as far away as the southern tip of South America.

Another case involved two Adelie Penguins that were transported 2,400 miles (3800 km) from their nesting grounds. The birds swam back home in 10 months, averaging half a mile (1 km) an hour.

Young birds use a combination of instinct and learned behavior to return to their wintering grounds, but some of them get lost along the way due to inexperience.

A Dutch researcher once trapped hundreds of European Starlings on their migration through the Netherlands and released them in Switzerland. The first-year birds departed immediately, but lost their way and ended up in Spain. Meanwhile, the experienced adults found their way to their wintering grounds in northern France, Belgium and southern England.

A 21,000-MILE (34,000-KM) ROUND TRIP

The most spectacular annual migration is that of the Arctic Tern. This species nests above the Arctic Circle in northern Canada. Traveling south at the end of the summer, it crosses the Atlantic, turns south along the coasts of Europe and Africa, to reach the Antarctic two months later. All told, it is a 15,000-mile (25,000-km) return trip—a trip around the world. Some individuals even make a little detour to take better advantage of the wind, in which case the total distance may be 21,000 miles (34,000 km).

The Greater Shearwater and Sooty Shearwater follow the reverse trajectory. They nest in the south, in New Zealand or the south Atlantic, and spend the southern winter in the north. Another shearwater, the Short-tailed, nests in southern Australia and spends the summer on the Alaskan coast. A South Polar Skua banded at the nest by researchers one southern summer was found six months later in Greenland, where it had died during the boreal summer.

Terns are not the only birds to make long journeys. Many other species achieve feats that are just as remarkable, given their size.

For example, many shorebirds that breed in the Canadian North cross North and South America each fall and spring, a voyage of more than 12,400 miles (20,000 km). There is a record of a Ruddy Turnstone, a handsome shorebird that nests in Scandinavia and on the Canadian tundra, leaving Alaska in August and reaching Hawaii three days later. It had traveled slightly over 620 miles (1000 km) a day at an average speed of 27 miles an hour (43 km/h).

TWENTY-FIVE-HUNDRED MILES (4000 KM) IN FOUR DAYS

The Blackpoll Warbler often leaves northern Canada on a direct 86-hour flight to northern South America. On arrival, little remains of the bird but feathers and bones. It has been calculated that if the warbler were to use gasoline as a fuel for its "motor," its "fuel efficiency" would be about 288,000 km a liter. The energy expenditure involved is equivalent to a man running a little more than 4 miles (6 km) a minute for 80 hours straight—an incredible feat.

Before starting south, the tiny Ruby-throated Hummingbird doubles its weight from one-tenth of an ounce (3 g) to one-fifth of an ounce. It leaves Canada in late summer, bound for northern

South America, a trip that entails a 560-mile (900 km) crossing of the Gulf of Mexico.

In Europe, the Northern Wheatear leaves Scandinavia for the British Isles, crossing 1,240 to 1,860 miles (2000 or 3000 km) of ocean to get there. Several European migratory species make an unbroken 680-mile (1100 km) trip across the Mediterranean. On completing the crossing, they rest awhile, before facing the 1,000-mile (1600 km) leg over the Sahara Desert.

Some falcons that nest in Asia have to fly 2,500 miles (4000 km) across the Indian Ocean to reach the East African coast, a three- or four-day trip.

A FLY-BY-NIGHT OPERATION

Migration date, flight speed, distance and destination are specific to each species. Some birds flock; others are relatively solitary. Many shorebirds travel only at night, but ducks and geese migrate around the clock.

The nocturnal migratory habits of many small passerines enable them to avoid predators. The small birds rest and feed by day, then resume their journey at sundown.

When food becomes scarce, normally sedentary Arctic species like the Snowy Owl will travel southward, often as far as Montreal or New York, and sometimes even farther.

Within a single species, some birds may travel south only a short distance; others are completely sedentary. In Canada, many Blue Jays, European Starlings and American Robins winter near the American border; others stay on their summer territory and are regular feeder visitors.

In Europe, some Blackbirds travel hundreds of miles to reach their wintering grounds; others never leave their territory. Which pattern they choose depends on what they have learned from their parents.

In the winter of 1994–95, hundreds of American Robins stayed in Canada instead of migrating south. Food was abundant, since wild fruit trees had produced an exceptional harvest. Those that successfully braved the winter temperatures (down to −13°F [−25°C] on a few occasions) found themselves well positioned to choose the best nesting sites in spring.

Other migratory species travel only a few hundred miles south to avoid periods of frost. During a cold snap, American Crows nesting in Canada migrate about 125 to 185 miles (200 to 300 km) south to the United States. They fly north as soon as the warm weather returns.

Live Long and Prosper

In general, birds lead a tough life that lasts only two to five years. Theirs is a constant struggle to find food and reproduce—which makes it easy to see why many of them die very young.

In some species, two-thirds of the fledglings die during their first year of life. It is estimated that 25 percent of those that remain die before two years of age. A study in Arizona showed that only 11 percent of Yellow-eyed Juncos reappeared at their nesting grounds the year after their birth.

Predators cause much of the mortality of young birds. Each year in Europe, European Sparrowhawks alone capture 18 to 34 percent of fledgling Great Tits. In North America, 70 percent of young Mourning Doves die before their first birthday, and the annual mortality rate for adults is 55 percent. Nevertheless, some individuals live as long as 10 years. As mentioned, the Mourning Dove is the most hunted bird on the North American continent.

Fatal accidents are frequent. Hummingbirds and sparrows are often caught in large spider webs. Barn Swallows can die of hunger when they become entangled in the tall grass they use to build their nests. Just as deadly are automobiles. In North America, nearly 60 million birds are killed each year in collisions with cars. Considerable mortality also results from birds fracturing their skulls against the plate-glass windows of office buildings and homes.

A VENERABLE DUCK

As a rule, small birds live shorter lives than large ones, yet some songbirds have stretched their physical and biological limits, setting records for longevity. One Ruby-throated Hummingbird, the only hummingbird species regularly found in northeastern North America, lived five years in the wild. The scientific literature also mentions the case of a male that fed in an Arizona garden for 14 years.

The following are age records in nature for a sampling of species: Great Spotted Woodpecker, 11; European Robin, Common Cuckoo and Evening Grosbeak, 13; Magpie and Northern Cardinal, 15; Lapwing and Canada Goose, 23; Herring Gull, 27; Golden Eagle, 36; Laysan Albatross, 53; and Common Black-headed Gull, 63.

One banded Mallard, a species highly prized by North American hunters, had reached the age of 29 when its carcass was found. An Arctic Tern lived 34 years—a long life, indeed, for the holder of the long-distance migration record. A more common lifespan for this species is only about 20 years.

A Centenarian in Captivity

Birds raised in captivity often live much longer than their relatives in the wild, due to constant protection and care. In Europe, a Garden Warbler lived to 23 in an aviary, far longer than the record of seven years for the species in nature.

One Rock Dove reached the venerable age of 32 in captivity, while a Canada Goose survived to 42. Some large cage parrots have become octogenarians. A Siberian Crane led a peaceful existence for 61 years at the Washington Zoo. A Graylag Goose survived to 26 in captivity, and a Golden Eagle lived to 46.

One Rock Dove lived 32 years in captivity.

One raven lived 69 years in human company, one year more than a captive Eagle Owl. Although disputed, the record may be 100, reportedly the age of an American Great Horned Owl raised in England.

To Flee or Not to Flee

Although flight is the best defense for many birds, some become accustomed to the presence of humans, and even seek them out. The reason is very simple: where there are humans, there is likely to be food. Huge agglomerations of gulls in ports and flocks of pigeons in public squares are familiar examples of this phenomenon.

In the Galápagos Islands, where birds have been isolated from humans for millennia, visitors can move about among their colonies without ruffling any feathers. On these isolated Pacific islands, small birds will even perch on a person's shoulder. Usually shy raptors seem unbothered, as do sea lions and iguanas. The rare human visitor to the equally isolated Antarctic can expect similar behavior from the resident penguins.

Birds that act tame in one environment may behave differently elsewhere. For example, the American Robin walks around on lawns, searching for worms, without much concern for human presence. In the forest, however, it is much more cautious and discreet. In England, the European Robin (which is not, like its American namesake, a thrush) frequents gardens. On the Continent it shuns humans, preferring forested habitats.

Some very tame birds, such as the Black-capped Chickadee and the Common Redpoll, can be enticed with a little patience to eat out of a person's hand. The Gray Jay, which ranges throughout the boreal forest, frequents human encampments in search of food. Even at its first contact with humans, it may perch on a person's hand or enter a cabin to pick up crumbs.

Unusual Behavior

In general, the wilder the habitat, the less likely birds are to be shy toward humans. One ornithologist visiting the North was able to pick kinglets out of trees as if they were small fruit. Tengmalm's

Owl, about 8 inches (20 cm) long, will occasionally allow itself to be caught in the same way.

In the 1930s, hundreds of Bohemian Waxwings invaded the University of Washington campus. They flew in circles around passersby, sometimes landing on them. In 1924, a young woman on a camping trip in New York State was surprised to see a pair of House Wrens enter her tent. The birds, which were nesting in a nearby tree, got in the habit of coming into the tent. One morning, they boldly perched on her pillow and tried to pull out some strands of her hair for their nest.

Some birds exhibit even stranger behavior. On a rainy fall day in Iowa, a doctor brought home a hummingbird to feed it. When it was hungry, the bird would call out. If its benefactor did not react quickly enough, the bird would go to his office, hover near his face or land on his hand, and stick out its tongue as if to drink nectar from a flower. The hummingbird was released 10 days later, when the weather got warm, so that it could continue its journey southward.

Some Ruby-throated Hummingbirds are so tame that they will take nectar from feeders only a few inches away from a human observer. Others will occasionally poke about in people's ears, no doubt under the impression that they are in the presence of new and remarkable varieties of flowers.

Bohemian Waxwings are sometimes extremely tame.

Some years ago in May, in the Laurentian Mountains in Canada, a man by the name of Gaétan Charbonneau, who was out for a walk, spotted a Ruffed Grouse behaving oddly in the forest. This species tends to be wary of humans, since it is a favorite of hunters, but that day the bird let Charbonneau approach to within 23 feet (7 m) without flying away. Bird and man met regularly after that. On the fourth occasion, the grouse readily perched on the hand of his new companion to eat a bit of bread. Eventually, all Charbonneau had to do was whistle in the forest and the bird would come out of hiding.

Although the reasons for this familiarity are unknown, there have been several similar reports about tame Ruffed Grouse over the past few decades.

Some birds seem completely unbothered by traffic or the noise of farm machinery. In other cases, they perceive vehicles as threats and attack them. Every morning for two years, a Greater Prairie Chicken living near a Minnesota airport would attack airplanes as they came in for a landing. As might be expected, the story has an unhappy ending: the bird died after colliding with a plane.

IMPRINTING

In some cases, birds raised by humans come to consider them their congeners, to the point where they follow them all over. Famed ethologist Konrad Lorenz studied this phenomenon, known as *imprinting*, at length.

Imprinting is a quick—indeed, almost instantaneous—form of learning that occurs in newborn chicks soon after hatching. In general, what is learned during imprinting will not be forgotten later. This phenomenon enables the bird to recognize its parents and members of its species. It also enables the chick to learn the basics of its song and how to produce it.

In the hours after they hatch, duck and goose chicks react to the beings in their immediate surroundings: in the absence of their real parents, they will gravitate toward whoever is feeding them. Birds raised in captivity, for example, will follow their adoptive (human) "parents" everywhere, at times even attempting to mate with them.

JUST FOR THE FUN OF IT

Ornithology is full of surprises, so could birds play the way young mammals do? As it happens, activities observed in certain species appear to be games.

In Iceland, observers have seen Common Eiders "shoot the rapids" of a river, then climb up the bank, return to the start and repeat the experience—just for the fun of it, so it seems. Groups of Adelie Penguins will use sloped ice as a "sliding pond." The Common Raven appears to take great pleasure in dropping twigs and catching them in midair.

Young corvids, including crows, ravens, choughs and magpies, pass objects back and forth in an elaborate ritual whose significance is unclear. Young raptors sometimes appear to play with objects as if they were hunting prey. It is believed that these "games" have a role in learning.

Passerines, especially House Sparrows, have been observed to toss pebbles onto a slanted roof, a sheet of glass or the ground. After analyzing this phenomenon, researchers concluded that the only logical explanation for the behavior was to hear the sound produced.

Common Eiders have been known to "shoot the rapids"— apparently just for the fun of it.

An Unhappy Ending

In recent years, considerable sums of money and energy have been devoted to protecting endangered species. The Peregrine Falcon in eastern Canada and the United States and the Capercaillie in Scotland are examples of two species that have been reintroduced.

Bird conservation fits within the bigger picture of protection of the environment. Unfortunately, the battle is far from being won. Commercial hunting may have been the main factor in the disappearance of species over the past two centuries, but this is no longer true today. Pollution and habitat destruction have become major problems.

A study of species extinction from the early seventeenth century to 1980 indicates that 90 percent of extinct birds were island dwellers. Island species are particularly vulnerable because of the rarity and precariousness of their habitat. In almost all cases, the cause of their demise was the introduction of competing species, combined with habitat destruction and excessive commercial hunting.

In our day, major disappearances of bird species are rare, but they receive little comment. Hardly a word was said about the extinction of the Uluguru Bushshrike of Tanzania in 1961; or of Semper's Warbler of Saint Lucia, seen for the last time in 1972. Several Socorro Doves, a Mexican species, still survive in captivity, but they have been absent from their natural environment for several years.

In 1987, Canadian and U.S. newspapers briefly reported the extinction of the Dusky Seaside Sparrow, the last five males of which had been kept in captivity for seven years. The demise of the species, once found only in the marshes around Cape Canaveral, Florida, had been hastened by the expansion of the Kennedy Space Center's facilities, the building of a new highway and a brush fire.

The largest birds that ever lived on the planet were the moas of New Zealand and the elephant birds of Madagascar. The moas, 10-feet (3-m)-tall ratites similar to the emus, vanished in the last century, apparently hunted out by the Maoris. The elephant birds

The Dodo of Mauritius was exterminated in the seventeenth century.

were even taller and weighed almost a thousand (450 kg) pounds. The volume of an elephant bird egg was 9 litres and each egg weighed up to 26 pounds or 13 kg (the average for an ostrich egg is 3 pounds or 1.5 kg). In fact, elephant bird eggs are considered the largest living cells the animal world has ever known.

The Dodo appears among the fictional characters in the children's book *Alice's Adventures in Wonderland*. But this turkey-sized bird actually existed on the island of Mauritius in the Indian Ocean. Sometimes described as an immense flightless pigeon, it was exterminated at the start of the seventeenth century, and was last sighted in 1662.

REQUIEM FOR THE GREAT AUK

All recent massive extinctions of birds have occurred in North America. The Labrador Duck, about which very little is known, became extinct in 1875. Although it was hunted commercially, experts believe it was never particularly abundant.

The Heath Hen had already existed for 100 years only in Martha's Vineyard, off the Massachusetts coast, when the last specimens succumbed to illness or hunting. Only 54 birds remained in 1924, and the last was seen in 1932.

The last Carolina Parakeet died in a zoo in 1918, but it had already been eradicated from the American landscape 10 years earlier because of the damage it caused to corn and orchard crops. It tore up apples to get at the seeds.

The Great Auk used to live in the Saint Lawrence estuary in northeastern Canada, in Greenland and on certain Icelandic and British islands. It measured 32 inches (80 cm) and stood erect on two feet like its distant South Pole cousins, the penguins.

Unfortunately, the Great Auk was unafraid of humans. It did not fly but could travel great distances in the water during migration. It wintered as far away as Florida and the coast of Spain. European explorers killed millions of Great Auks for supplies of fresh meat, and fishermen used them for bait. The largest known colony in the eighteenth century was on Funk Island, off

The Carolina Parakeet was eradicated from the planet because of its habit of feeding on commercial crops.

THE SECRET LIVES OF BIRDS

the coast of Newfoundland. The decline of auks accelerated when people began to use their feathers for beading, and their oil. The colony was entirely wiped out between 1585 and 1841. The last two specimens were killed on June 4, 1844, near an island off the coast of Iceland.

THE EPIC OF THE PASSENGER PIGEON

Of all modern extinctions, that of the Passenger Pigeon is the most spectacular and best documented.

For decades, the Passenger Pigeon seems to have been the most abundant species in North America. Estimates vary from three to five billion individuals, or 25 to 40 percent of the bird population of the United States.

The birds used to migrate in such numbers that they would darken the sky. One contemporary eyewitness described a spring flock that formed a ribbon 300 miles (500 km) long and about half a mile (1 km) wide. The birds' droppings, he said, fell from the sky like a wet snowstorm.

According to numerous observers cited by Arthur Cleveland

The Great Auk was hunted to extinction for its meat and feathers.

Bent in *Life Histories of North American Birds*, the pigeon flocks produced a muffled sound similar to the rumbling of Niagara Falls, or even a tornado.

During migration, the forests in which the birds roosted were literally devastated. The birds left behind a layer of droppings several inches thick. Many trees of up to 23 inches (60 cm) in diameter were broken under the weight of so many birds. The highest branches fell as if the forest had been struck by a tornado.

THE LAST OF THE PIGEONS

The Passenger Pigeon was mainly a tree-dwelling species that nested east of the American and Canadian Great Plains, usually in beech forests. Its nesting grounds were as large as 300 square miles (500 km²) and in these forests, almost all branches were covered by nests. It was not uncommon to find up to 100 in a single tree.

According to observers at that time, the Passenger Pigeon's nest was similar to the Mourning Dove's: a rickety mass of twigs usually at least 10 feet (3 m) above the ground. Each nest commonly contained two eggs. Considering the population density, these were possibly laid by two different females. Ornithologists disagree about the number of broods raised per year. Guesses range from a single brood per pair to three or four every year.

The eggs were incubated for 14 days by both parents. Newly hatched chicks were fed for two more weeks on food regurgitated by the adults, then left to their own devices. After surviving for a few days on their fat reserves, the young would leave the nest.

The Passenger Pigeon's diet consisted of beechnuts, acorns,

chestnuts, berries and insects. It was particularly fond of salt. It fact, farmers frequently used salt as bait for the bird, since they considered the species a major pest.

The Passenger Pigeon first appears in the literature about 1630 (the reference is to millions and millions of birds) and the start of its decline dates back to 1870. The last specimen in the wild was killed in 1898. Several years earlier, merchants had already grown concerned about the bird's rarity. What happened?

FIVE THOUSAND HUNTERS

Two factors were responsible for the disappearance of the Passenger Pigeon: commercial hunting and the transformation of most of its nesting territory into agricultural land.

Bent reports that, in 1879, 5,000 people in the United States were hunting the pigeon commercially all year long both at its nesting sites and on its wintering grounds in the southern states. The hunters used nets that could capture about 2,500 birds a day, sometimes even up to 5,000. The birds were delivered to the city by boat or railroad, but many were captured alive, to be used as targets on shooting ranges.

The birds were also exploited for home use. Several days before the parent birds left the nest, entire villages would meet on nesting sites to capture the juveniles. In some cases, trees with a large number of nests would be felled in order to capture 200 birds at one stroke.

Quebeckers, great lovers of pigeon meat, even shot them from out of the windows of their urban homes. Quebec City authorities of the time went so far as to adopt a regulation prohibiting hunting within city limits because the gun wads would sometimes set fire to the roofs of houses.

The flesh of the Passenger Pigeon was considered a delicacy. It was served fricasseed, and young birds were spit-roasted. Many birds were captured and fattened for culinary purposes.

The Passenger Pigeon, a species that once darkened American skies by its sheer numbers, was wiped out in only a few decades. The last pigeon lived in the Cincinnati Zoo. Called Martha, she was born in captivity at a time when all the wild members of her species had already been eliminated for more than decade. Martha died at the ripe old age of 29 on September 1, 1914, at 1:00 p.m.

Epilogue

Over the past century, the world of birds has seen some shake-ups and the populations of several species have fluctuated markedly. Herring Gulls in Europe and Ring-billed Gulls in North America are much more numerous than in the past because of the abundant food supply they find in garbage dumps on the outskirts of our cities. Conversely, the Barn Swallow is becoming rarer on both continents due to a decrease in suitable nesting sites.

The Canada Goose population that nests in the Canadian North is also declining, in this case due to hunting, despite government-imposed bag limits. However, the goose populations introduced into a number of American and Canadian cities several decades ago are exploding, and municipal authorities are doing what they can to rein the geese in. Because of their urban habits, these sedentary birds are not sought after by hunters.

At the turn of the century, the Snow Goose population in the

Atlantic migratory flyway numbered about 4,000. Today, despite hunting, it has increased to 600,000. Biologists believe that the population will decline over the next few years because its food, mainly grasses, has become scarcer on its Arctic nesting grounds.

Although bird populations fluctuate considerably over time, it is to be hoped that today's strict hunting regulations will prevent a repeat of the tragedy of the Passenger Pigeon. In the Province of Quebec, for example, the Harlequin Duck is off-limits to hunters because of its rarity. Environmental disturbance, however, remains the chief threat to the survival of birds today.

In North America, hunters have spent nearly $500 million to protect the wetlands favored as breeding grounds by ducks and many other birds. As early as the 1930s in the Canadian West, wetlands were being re-created after prolonged droughts in an effort to save the ducks. By protecting habitats, hunters' associations and environmental groups have helped maintain water bird populations.

For their part, thousands of bird observers around the world help gather scientific data on nesting, distribution and the presence of rare species. Through their vigilance, birds will continue to reveal their secrets to us for a long time to come.

The Harlequin Duck is one of the most colorful ducks in North America.

Bibliography

Bellerose, Frank C. *Ducks, Geese and Swans of North America.* Mechanicsburg, PA: Stackpole Books, 1980.

Bent, Arthur Cleveland. *Life Histories of North American Jays, Crows, and Titmice.* New York: Dover Publications, 1988.

———. *Life Histories of North American Birds of Prey.* Vol. 1. New York: Dover Publications, 1961.

———. *Life Histories of North American Blackbirds, Orioles, Tanagers and Allies.* New York: Dover Publications, 1965.

———. *Life Histories of North American Gallinaceous Birds.* New York: Dover Publications, 1963.

———. *Life Histories of North American Wagtails, Shrikes, Vireos and Their Allies.* New York: Dover Publications, 1965.

British Ornithologists' Union. *A Dictionary of Birds.* London: Vermilion, 1985.

Cadman, Michael D., et al. *Atlas of the Breeding Birds of Ontario.* Waterloo: University of Waterloo Press, 1987.

Dorst, Jean. *The Life of Birds.* Trans. I.C.J. Galbraith. New York: Columbia University Press, 1974.

Génsbol, Benny. *Guide to the Birds of Prey of Britain and Europe, North Africa and the Middle East.* Trans. Gwynne Vevers. Toronto: Collins, 1984.

Gill, Frank B. *Ornithology.* New York: W.H. Freeman and Company, 1994.

Gingras, Pierre. "A tire d'aile." *La Presse,* 1987–1995.

Godfrey, W. Earl. *The Birds of Canada*. Revised ed. Ottawa: National Museum of Natural Sciences, 1986.

Harrison, Hal H. *A Field Guide to Birds' Nests*. Boston: Houghton Mifflin, 1975.

Leahy, Christopher. *The Birdwatcher's Companion*. New York: Bonanza Books, 1982.

Lockley, Ronald M. *Seabirds of the World*. New York: Facts on File, 1983.

Madge, Steve, and Hilary Burn. *Waterfowl: An Identification Guide to the Ducks, Geese, and Swans of the World*. Boston: Houghton Mifflin, 1988.

Monroe, Burt L., and Charles G. Sibley. *A World Checklist of Birds*. New Haven, CT: Yale University Press, 1993.

National Geographic. *Field Guide to the Birds of North America*. Washington, D.C.: National Geographic, 1987.

Perrins, Christopher M., and Alex L.A. Middleton. *The Encyclopedia of Birds*. New York: Facts on File, 1985.

Peterson, Roger Tory. *A Field Guide to the Birds East of the Rockies*. Boston: Houghton Mifflin, 1980.

Schneider, Dan. "Starling Wars." *Nature Canada* (Fall 1990).

Stokes, Donald W. *A Guide to Bird Behavior*. Vol. 1. Boston: Little, Brown, 1979.

Stokes, Donald W., and Lilian Stokes. *A Guide to Bird Behavior*. Vol. 2. Boston: Little, Brown, 1983.

———. *A Guide to Bird Behavior*. Vol. 3. Boston: Little, Brown, 1989.

Terres, John K. *The Audubon Society Encyclopedia of North American Birds*. New York: Alfred A. Knopf, 1982.

Index

THE SECRET LIVES OF BIRDS